NELSON
VICmaths

VCE UNITS ③ + ④

mathematical
methods 12

mastery workbook

Greg Neal
Sue Garner
George Dimitriadis
Toudi Kouris
Stephen Swift

Nelson VICmaths Mathematical Methods 12 Mastery Workbook
1st Edition
Greg Neal
Sue Garner
George Dimitriadis
Toudi Kouris
Stephen Swift
ISBN 9780170464062

Publisher: Dirk Strasser
Additional content created by: ansrsource
Project editor: Alan Stewart
Series cover design: Leigh Ashforth (Watershed Art & Design)
Series text design: Rina Gargano (Alba Design)
Series designer: Nikita Bansal
Production controller: Karen Young
Typeset by: MPS Limited

Any URLs contained in this publication were checked for currency during the
production process. Note, however, that the publisher cannot vouch for the
ongoing currency of URLs.

Acknowledgements
TI-Nspire: Images used with permission by Texas Instruments, Inc
Casio ClassPad: Shriro Australia Pty. Ltd.

For product information and technology assistance,
in Australia call **1300 790 853**;
in New Zealand call **0800 449 725**

For permission to use material from this text or product, please email
aust.permissions@cengage.com

ISBN 978 0 17 046406 2

Cengage Learning Australia
Level 7, 80 Dorcas Street
South Melbourne, Victoria Australia 3205

Cengage Learning New Zealand
Unit 4B Rosedale Office Park
331 Rosedale Road, Albany, North Shore 0632, NZ

For learning solutions, visit **cengage.com.au**

Printed in China by 1010 Printing International Limited.
2 3 4 5 6 7 26 25 24 23

Contents

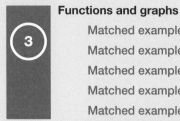

Functions and transformations

Applications of differentiation

Discrete probability and the binomial distribution   188

To the student

Nelson VICmaths is your best friend when it comes to studying Mathematical Methods in Year 12. It has been written to help you maximise your learning and success this year. Every explanation, every exam hack and every worked example has been written with the exams in mind.

Here are the 3 steps to mastering each topic

STEP 1

Study every Worked Example

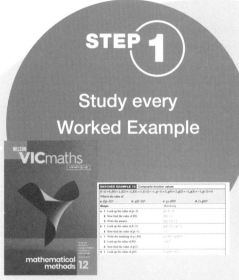

The 3 steps to mastering each topic

STEP 2

Complete the Matched Example in the Mastery Workbook

STEP 3

Do the Mastery questions in the exercise that are linked to the Worked Example

CHAPTER

POLYNOMIALS

(1)

MATCHED EXAMPLE 1 | Polynomial expressions and their features

SB
p. 4

Determine whether each expression is a polynomial, giving reasons if it is not. State the degree, leading term and coefficients of any polynomial found.

a $1 + \dfrac{1}{x} + \dfrac{1}{x^2} + \dfrac{1}{x^3}$

b $x^0 + \dfrac{1}{2}x^2 + \left(\dfrac{x}{3}\right)^3 - 3^{\frac{1}{2}}x^4$

Steps	Working
a Write the expression with descending powers. Not all of the powers are positive whole numbers, so it is not a polynomial.	
b 1 Write the expression with descending powers. All of the powers are positive whole numbers, so the expression is a polynomial. 2 Identify the degree, leading term and coefficients.	

SB

p. 5

MATCHED EXAMPLE 2 | Simplifying polynomials

If $f(x) = x^3 + 3x^2 + 5x$ and $g(x) = 6x^2 + 5x + 3$, find:

a $2f(x) - g(x) + 2$ and state its degree

b $f(x)g(x)$ and state its degree

Steps	Working
a **1** Substitute for $f(x)$ and $g(x)$.	
2 Add or subtract like terms to simplify.	
b **1** Substitute for $f(x)$ and $g(x)$.	
2 Multiply using the index law $a^m a^n = a^{m+n}$. Then add or subtract like terms to simplify.	

MATCHED EXAMPLE 3 | Substituting values into polynomials

SB
p. 5

If $Q(x) = mx^4 - 2x^3 + 6x - 6$ and $Q(2) = 6$, obtain a value for m.

Steps	Working
1 Substitute $(2, 6)$ into $Q(x)$.	
2 Solve for m.	

Using CAS 1:
Defining and
evaluating
polynomials
p. 5

MATCHED EXAMPLE 4 | Modelling polynomial functions

The relationship between the height of a plant and the temperature of the greenhouse in which the plant is in is modelled by the equation $h(t) = -0.125t^2 + 6t - h_0$, where h is the height of the plant, in cm, t is the temperature of the greenhouse, in °C, and h_0 is the height of the plant at 0°C.

If height of the plant is 40 cm at 0°C,

a find the height of the plant at temperature t°C

b determine the height of the plant grown at a temperature of 10°C

c determine the temperature of the greenhouse at which the plant height will be 0 cm

Steps	Working
a Substitute $h_0 = 40$ into $h(t)$.	
b Substitute $t = 10$ into $h(t)$ and simplify by collecting like terms.	
c 1 Substitute $h(t) = 0$.	
2 Check the feasibility of the solutions.	

MATCHED EXAMPLE 5	Equating coefficients

Let $M(x) = x^4 + 2x^3 + qx^2 + 3x + p$ and $N(x) = 2x^2 - xq + 3$.

a Show that $2M(x) - 4N(x) = 2x^4 + 4x^3 + (2q - 8)x^2 + (6 + 4q)x + 2p - 12$.

b If $2M(x) - 4N(x) = 2x^4 + 4x^3 + 22x^2 - 8$, find the values of p and q for $x \in R$.

SB

p. 8

Steps	Working
a **1** Evaluate $2M(x)$.	
2 Evaluate $4N(x)$.	
3 Evaluate $2M(x) - 4N(x)$.	
b **1** Equate the coefficients of the x^2 term and solve for q.	
2 Equate the coefficients of the independent of x term and solve for p.	

MATCHED EXAMPLE 6 | Long division

Divide $P(x) = x^3 - 10x^2 + 30x + 6$ by $d(x) = x - 5$ and hence express $M(x)$ in the form $d(x) \times Q(x) + R$.

Steps	Working
1 Divide the highest power of x from $(x - 5)$ into the highest power of $(x^3 - 10x^2 + 30x + 6)$. This is $x^3 \div x$, which gives x^2. Write this up the top in the x^2 column. Work out how much of $(x^3 - 10x^2)$ has been accounted for by doing $(x - 5) \times x^2 = x^3 - 5x^2$. Write this under the $(x^3 - 10x^2)$ and subtract to work out the x^2 remaining $(-5x^2)$. Then bring down the $+ 30x$.	
2 x into $-5x^2$ goes $-5x$ times. Write the $-5x$ in the x column, above the $+30x$. $(x - 5) \times (-5x) = -5x^2 + 25x$. Put this underneath the $-5x^2 + 30x$ and subtract to get $5x$. Bring down the 6.	
3 x goes into $5x$ five times. Put the 5 above the 6. $(x - 5) \times 5 = 5x - 25$. Put this underneath the $5x + 6$ and subtract to get 31.	
4 Write the answer in the form $P(x) = d(x) \times Q(x) + R$	

MATCHED EXAMPLE 7 | Division by inspection

Given $M(x) = 2x^2 + 5x + 4$ and $d(x) = x + 1$, express $\dfrac{M(x)}{d(x)}$ in the form $N(x) + \dfrac{R}{d(x)}$.

Steps	Working
1 Write the numerator $2x^2 + 5x + 4$ as $2x(x + 1) + 3(x + 1) + 1$.	
2 Factorise the first two terms of the numerator.	
3 Split the expression into two partial fractions.	
4 Simplify.	

1

MATCHED EXAMPLE 8 Division of a polynomial by a non-linear divisor

Divide $P(x) = x^3 - 6x^2 + 12x + 6$ by $d(x) = -x^2 - 2$ to find the quotient and remainder.

Steps	Working

1 Divide the highest power of x from $(-x^2 - 2)$ into the highest power of $(x^3 - 6x^2 + 12x + 6)$.

This is $x^3 \div (-x^2)$, which gives $-x$.

Put this up the top in the x place.

Work out how much of $(x^3 - 6x^2)$ has been accounted for by doing $(-x^2 - 2) \times (-x) = x^3 + 2x$.

Put this under the $(x^3 - 6x^2)$ and subtract to work out how many x^2 and x are left.

Then bring down the 6.

2 $-x^2$ into $-6x^2$ goes 6 times. Put the 6 in the units place, above the 6.
$(-x^2 - 2)(6) = -6x^2 - 12$

Put this underneath the $-6x^2 + 10x + 6$ and subtract to get $10x + 18$.

3 Write the answer.

MATCHED EXAMPLE 9 Finding the remainder

Find the remainder when $2x^3 - 2x^2 + 4x + 3$ is divided by $2x - 1$.

Steps	Working
1 Write down $P(x)$.	
2 The remainder when $P(x)$ is divided by $2x - 1$ is $P\left(\dfrac{1}{2}\right)$. Evaluate $P\left(\dfrac{1}{2}\right)$.	
3 Write the answer.	

MATCHED EXAMPLE 10 | Using the remainder theorem to find a term coefficient

Evaluate k if the remainder is -8 when dividing $x^3 - kx^2 + 3x + 1$ by $x - 3$.

Steps	Working
1 Write down $P(x)$.	
2 The remainder when $P(x)$ is divided by $x - 3$ is $P(3)$.	
3 The remainder is -8. So, $P(3) = -8$.	
4 Solve the equation for k.	

MATCHED EXAMPLE 11 Using the remainder theorem to find unknowns

$f(x) = 2x^3 - ax^2 + bx + 3$, where a, b are constants.

a Given that when $f(x)$ is divided by $(x - 1)$ the remainder is 3, show that $a - b = 2$.

b Given also that when $f(x)$ is divided by $(x - 3)$ the remainder is 27, find the value of a and b and hence determine the polynomial $f(x)$.

Steps	Working
a 1 Write down $f(x)$.	
2 The remainder when $f(x)$ is divided by $x - 1$ is $f(1)$.	
3 The remainder is 3. So, $f(1) = 3$.	
4 Simplify the equation.	
b 1 The remainder when $f(x)$ is divided by $x - 3$ is $f(1)$.	
2 The remainder is 27. So, $f(3) = 27$.	
3 Set up a pair of simultaneous equations.	
4 Solve for a and b using simultaneous equations.	
5 Write the answer.	

SB
Using CAS 3:
Finding the
remainder of a
polynomial
p. 16

p. 17

MATCHED EXAMPLE 12 | Factorising polynomials using long division

Factorise $f(x) = x^3 - 5x^2 + 2x + 8$.

Steps	Working
1 Write the polynomial.	
2 Try factors of 8 (i.e., $\pm 1, \pm 2, \pm 4$) until the remainder of 0 is found.	
3 Use the process of polynomial long division to divide $x + 1$ into $f(x)$.	
4 Factorise $x^2 - 6x + 8$.	
5 State your answer.	

9780170464062

MATCHED EXAMPLE 13	Factorising polynomials using synthetic division

Factorise $f(x) = x^4 + 5x^3 + 2x^2 - 20x - 24$.

Steps	Working

1 Write the polynomial.

2 Try factors of 24 (i.e., ±1, ±2, ±3, ±4, ±6, ±8, ±12, ±24) until the remainder of 0 is found.

3 Place the 2 in the top left-hand corner of the first row, to the left of the separation line.

    Bring the first coefficient, 1, down.

    Place the coefficients of the terms in $f(x)$ in the top row to the right of the separation line, in order of their decreasing power.

    The values in the second row are found by multiplying each value in the bottom row by the 2 from $(x - 2)$. The values in the bottom row are column totals, which make up the coefficients of the quotient, the last value being the remainder.

4 Write the answer in the form $f(x) = d(x) \times g(x) + R$.

5 Write down the quotient.

6 To factorise the quotient, repeat the process by first finding a linear factor of $g(x)$. Try factors of 12 (i.e., ±1, ±2, ±3, ±4, ±6, ±12) until the remainder of 0 is found.

    The rest of the values in the second row are found by multiplying each value in the bottom row by the −2 from $(x + 2)$. The values in the bottom row are column totals, which make up the coefficients of the quotient, the last value being the remainder.

7 Write down the factorised form of $f(x)$.

8 Continue factorising by factorising the quadratic polynomial.

p. 18

MATCHED EXAMPLE 14 | Factorising polynomials by equating coefficients

Factorise $f(x) = 2x^3 - 5x^2 + x + 2$.

Steps	Working
1 Write the polynomial.	
2 Try factors of 2 (i.e., ±1, ±2).	
3 Write $f(x)$ in the form of $(x - a)Q(x)$, where $Q(x)$ is a quadratic written in general form and a, b and c are constant coefficients.	
4 Find a by equating coefficients of x^3.	
5 Find c by equating the constant terms.	
6 Find b by equating coefficients of x^2. Note: b can also be found by equating coefficients of x.	
7 State your answer in the form $P(x) = (x - a)Q(x)$.	
8 Factorise completely.	

Using CAS 4:
Finding factors of
polynomials
p. 19

Using CAS 5:
Finding the
real factors of
polynomials
p. 19

MATCHED EXAMPLE 15	Factorising using the sum and differences of two cubes

SB

p. 20

Factorise each expression.

a $27y^3 - 1^3$

b $(x + 1)^3 + 2^6$

Steps	Working
a 1 Recognise the difference of two cubes.	
2 Use the formula $a^3 - b^3 = (a - b)(a^2 + ab + b^2)$, where $a = 3y$ and $b = 1$.	
b 1 Recognise the sum of two cubes.	
2 Use the formula $a^3 + b^3 = (a + b)(a^2 - ab + b^2)$, where $a = x + 1$ and $b = 4$.	
3 Simplify.	

MATCHED EXAMPLE 16 | Solving equations involving perfect cubes

Solve $216 - 8(p + 2)^3 = 0$.

Steps	Working
1 Subtract 216 from both sides.	
2 Divide both sides by -8.	
3 Take the cube root of both sides.	
4 Subtract 2 from both sides.	

MATCHED EXAMPLE 17 | Solving quartic equations

Solve $2x^4 + x^3 - 2x^2 - x = 0$.

Steps	Working
1 Take x out as a common factor.	
2 Use the cubic polynomial and try factors of 2 (i.e., ± 1, ± 2).	
3 Method 1: Using long division Divide $P(x)$ by $x - 1$ by using long division.	
Method 2: Using synthetic division Divide $P(x)$ by $x - 1$ by using synthetic division.	
Method 3: By equating coefficients Divide $P(x)$ by $x - 1$ by equating coefficients. Write $P(x)$ in the form of $(x - a)Q(x)$, where $Q(x)$ is a quadratic written in general form $ax^2 + bx + c$, and a, b and c are constant coefficients.	
4 State the polynomial as a product of the factors found so far.	
5 Factorise the quadratic.	
6 Use the null factor law.	
7 Solve the equation.	

MATCHED EXAMPLE 18 | Sketching a cubic function of the form $y = a(x - h)^3 + k$

Sketch the graph of $y = 2(x - 1)^3 - 16$.

Steps	Working
1 Compare $y = 2(x - 1)^3 - 16$ with $y = 2x^3$.	
2 Find the x-intercept(s).	
3 Find the y-intercept.	
4 Sketch the graph. It will have the same shape as $y = 2x^3$.	

MATCHED EXAMPLE 19	Sketching a cubic function of the form $y = ax^3 + bx^2 + cx + d$

Sketch the graph of $y = x^3 - x^2 - 6x$, showing the coordinates of all intercepts.

Steps	Working
1 Factorise $y = x^3 - x^2 - 6x$.	
2 Find the x-intercepts.	
3 Find the y-intercept.	
4 Determine the direction of the graph.	
5 Sketch the graph.	

| MATCHED EXAMPLE 20 | Sketching a quartic function |

Sketch the graph of $y = -x^4 + 25x^2$, showing all intercepts.

Steps	Working
1 Factorise $y = -x^4 + 25x^2$.	
2 Find the x-intercepts.	
3 Find the y-intercept.	
4 Determine the direction of the graph.	
5 Sketch the graph.	

9780170464062

MATCHED EXAMPLE 21 Sketching a polynomial of a higher power

For $f(x) = -x^3(x-3)^4(x-2)^5$ state the leading term, the zeros and hence sketch the graph.

Steps	**Working**
1 State the leading term.	
2 State the zeros (x-intercepts).	
3 Find the y-intercept.	
4 Determine the direction of the graph.	
5 Sketch the graph.	

MATCHED EXAMPLE 22 | Determining the rule of a polynomial function from a graph

The rule for the function with the graph shown is of the form $y = a(x - h)^3 + k$. Find the values of a, h and k. Hence, state the equation of the function.

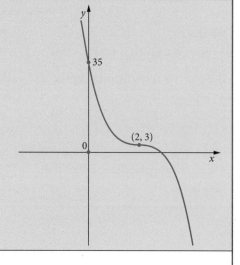

Steps	**Working**
1 Using the stationary point of inflection, find the value of h and k.	
2 Substitute $(0, 35)$ into the general equation.	
3 Write the equation of the function.	

MATCHED EXAMPLE 23 | Modelling problem

A miniature toy factory has been studying their profit for an upcoming production. The figure shows the graph of a cubic polynomial function used to model the profit achieved as a function of the number of toys sold. Use the graph to answer the questions below. Estimate values to the nearest one unit on the horizontal axis and to the nearest 1000 units on the vertical axis.

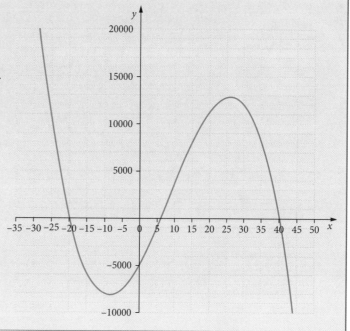

a How many toys must the factory sell to maximise profit? What is the maximum profit?

b What are the zeros of the function P?

c Find an equation to represent this function.

d State the domain for this model.

Steps	Working
a Look for the maximum on the graph.	
b State the zeros (x-intercepts).	
c 1 Write a factorised general model for this cubic function in terms of x. 2 Find the constant 'a' using another point, say (26, 13 000).	
d Restrict the domain so that the profit >0.	

CHAPTER

2 EQUATIONS

SB

p. 47

MATCHED EXAMPLE 1	Solving quadratic equations

Solve the equation $-3x^2 = 8x - 4$ for x.

Steps	Working
1 Rewrite the equation to equal 0.	
2 As the equation does not factorise, use the quadratic formula. Identify a, b and c.	
3 Substitute into the quadratic formula.	

MATCHED EXAMPLE 2	Using the discriminant

Find the value of k for which the graph of $y = kx - 3$ and $y = x^2 + 6x - 2$ do not intersect.

Steps	Working
1 Substitute $kx - 3$ for y in the second equation.	
2 Rewrite the equation equal to 0.	
3 If the graphs do not intersect, then this equation will have no solution, so $\Delta < 0$. Identify a, b and c.	
4 Find the value of $b^2 - 4ac$.	
5 Solve $b^2 - 4ac < 0$.	
6 Solve the inequality by graphing the discriminant and identifying where $\Delta < 0$ (below the horizontal axis), namely between $k = 4$ and $k = 8$.	
7 Answer the question.	

2

SB

Using CAS 1:
Solving
inequalities
p. 48

SB

Using CAS 2:
Determine the
point(s) of
intersection by
graphing
p. 49

MATCHED EXAMPLE 3	Finding values of a parameter for simultaneous linear equations, using the coefficients method

Find the values of k for which the simultaneous equations

$(k-8)x + 4y = k - 8$

$-3x + ky = 3$

have no solution.

Steps	Working

1 For the lines to have no solution they must have the same gradient.

$\therefore \dfrac{a}{c} = \dfrac{b}{d}$

Solve the equation to find two values of k.

2 The lines must have different equations.

$\dfrac{a}{c} = \dfrac{b}{d} \neq \dfrac{e}{f}$

Write the ratio of the coefficients of x and the ratio of the constants in terms of k.

These coefficients must have different values for the lines to have the same gradients but different y-intercepts.

Calculate the value of these ratios by substituting $k = 6$.

3 Calculate the value of the ratios by substituting $k = 2$.

SB

p. 55

| MATCHED EXAMPLE 4 | Determining parameter values that produce a unique solution |

Find the values of m for which the simultaneous equations

$$mx - 4y = 5$$

$$x + (m - 5)y = m + 2$$

have one unique solution.

Steps	Working
1 For the lines to have one solution, they must have different gradients. $\dfrac{a}{c} \neq \dfrac{b}{d}$	

MATCHED EXAMPLE 5 | Solving a trigonometric equation for a given domain

Solve $\tan(x) = \sqrt{3}$, $x \in [0, 2\pi]$ for x.

Steps	Working
1 Find the reference angle.	
Refer to the exact value table.	
2 Identify the quadrants where there are solutions.	
Find the quadrants where $\tan(x)$ is positive.	
3 Find the solutions.	
Find the equivalent angle to $\dfrac{\pi}{3}$ in quadrant 3.	

MATCHED EXAMPLE 6 | Solving a circular function for the angle *nx*

Solve $\cos(2x) = -\dfrac{1}{\sqrt{2}}, x \in [0, 2\pi]$ for *x*.

Steps	**Working**
1 Determine the reference angle.	
Refer to the exact value table.	
2 Identify the quadrants where there are solutions.	
Find the quadrants where $\cos(2x)$ is negative.	
3 Determine the domain for $2x$.	
If $x \in [0, 2\pi]$, then $2x \in [0, 4\pi]$.	
A domain of $[0, 4\pi]$ will produce 4 solutions.	
4 Determine the solutions for $2x$.	
$\dfrac{3\pi}{4}$ and $\dfrac{5\pi}{4}$ are in the first revolution.	
To obtain the next two solutions, add 2π or $\dfrac{8\pi}{4}$ to these values.	
5 Determine the solutions for *x*.	
Divide both sides by 2.	

MATCHED EXAMPLE 7	Solving trigonometric equations of the form sin (*nx*) = *m* cos (*nx*)

Solve $\sqrt{3}\sin(2x) - \cos(2x) = 0, 0 \le x \le 2\pi$ for x.

Steps	Working
1 Add cos (2*x*) on both sides and divide both sides by $\sqrt{3}$ cos (2*x*).	
2 Determine the reference angle. Refer to the exact value table.	
3 Identify the quadrants where tan (2*x*) is positive.	
4 Determine the domain for 2*x*.	
5 Solve the equation for 2*x*. Determine the solutions for 2*x*. $\dfrac{\pi}{6}$ and $\dfrac{7\pi}{6}$ are in the first revolution. To obtain the next two solutions, add 2π or $\dfrac{12\pi}{6}$ to these values.	
6 Solve the equation for *x*. Divide both sides by 2.	

2

MATCHED EXAMPLE 8 | Finding general solutions for trigonometric equations

Solve $\cos 4x = -\dfrac{1}{2}$ for x.

Steps	**Working**
1 Determine the reference angle. Refer to the exact value table.	
2 Identify the quadrants where $\cos(4x)$ is negative.	
3 Determine the first two solutions that are in the domain $[-\pi, \pi]$ and add $2n\pi$ to both these solutions	
4 Divide by 4 for the general solution for x.	

> There is an infinite number of solutions for x. Including '$n \in Z$' indicates n must be an integer.

MATCHED EXAMPLE 9 | Finding general solutions for tan (*x*)

Write the general solution for *x* that satisfies the equation $\tan(x) = \sqrt{3}$.

Steps	Working
1 Find the reference angle.	
2 Identify the quadrants where tan (*x*) is positive.	
3 Find one solution in the domain $[-\pi, \pi]$ and add $n\pi$	

For equations involving tan, add multiples of π, not 2π, because tan has a period of π.

MATCHED EXAMPLE 10 | Solving exponential equations

Solve $27^{x+1} = 9^{x+2} \times 81$ for x.

Steps	Working
1 Express each term as a power of 3.	
2 Expand the brackets and multiply the powers.	
3 Equate the powers.	
4 Solve the linear equation.	

MATCHED EXAMPLE 11	Solving quadratic exponential equations

Solve $2^{2x} - 6(2^x) + 8 = 0$ for x.

Steps	**Working**
1 Let a represent the linear term.	
2 Express the equation in terms of a.	
3 Factorise and solve for a.	
4 Rewrite the solutions in terms of x.	
5 Solve by writing all terms as a power of 2 and equating the powers.	

2

MATCHED EXAMPLE 12 | More quadratic exponential equations

Solve $e^{2x} - 2 = 4 - e^x$ for x.

p. 66

Steps	Working
1 Add 2 on both sides of the equation.	
2 Let a represent the exponential term.	
3 Express the equation in terms of a.	
4 Factorise and solve for a.	
5 Rewrite the solutions in terms of x.	
6 Solve by writing the equations in log form.	

9780170464062

MATCHED EXAMPLE 13 | Solving logarithmic equations 1

Solve $\log_2(12x+8)-\log_2(x)=4$ for x.

Steps	Working
1 Use log laws to express the left-hand side of the equation as a single log.	
2 Change the equation from log form to index form.	
3 Simplify and solve the equation.	

MATCHED EXAMPLE 14 | Solving logarithmic equations 2

Solve $\log_e(x)+\log_e(x+1)=\log_e 20$ for x.

Steps	Working
1 Use log laws to express the left-hand side of the equation as a single log.	
2 Equate the brackets and solve the quadratic equation.	
3 Check to see if both solutions are valid in the original equation.	
4 Write the solution.	

MATCHED EXAMPLE 15 | Solving literal equations

Solve $yp+3q=2p$ for y.

Steps	Working
1 Solve by inverse operations. Subtract $3q$ from both sides.	
2 Divide both sides by p.	

MATCHED EXAMPLE 16 Solving exponential literal equations

Solve for x when $2a^{-4x} - 4c + 6ky = 0$.

Steps	Working
1 Add $4c$ to both sides and subtract $6ky$ from both sides. Then divide both sides by 2.	
2 Change to log form.	
3 Divide both sides by -4.	

MATCHED EXAMPLE 17 | Solving logarithmic literal equations

Solve for x when $k = 2\log_a\left(\dfrac{x+b}{y}\right) - c$.

Steps	Working
1 Add c to both sides and divide both sides by 2. Then change to index form and swap sides so that x is on the left-hand side.	
2 Multiply both sides by y.	
3 Subtract b from both sides.	

Using CAS 5:
Solving literal
equations
p. 70

MATCHED EXAMPLE 18   Finding approximate solutions using Newton's method 1

Use Newton's method to approximate a zero for the cubic function $y = x^3 + 2x^2 - x - 2$. If $x_0 = 0$, find the value of x_1.

Steps	Working

1 Differentiate $y = x^3 + 2x^2 - x - 2$.

2 Calculate $f(x_0)$ and $f'(x_0)$ where $x_0 = 0$.

3 Use Newton method to find x_1.

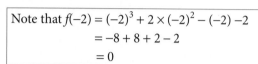

$$x_{n+1} = x_n - \frac{f(x_n)}{f'(x_n)}$$

$$\boxed{\begin{aligned} \text{Note that } f(-2) &= (-2)^3 + 2 \times (-2)^2 - (-2) - 2 \\ &= -8 + 8 + 2 - 2 \\ &= 0 \end{aligned}}$$

MATCHED EXAMPLE 19 | Finding approximate solutions using Newton's method 2

Complete two iterations of Newton's method to solve $x^3 + 2x + 1 = 0$, correct to four decimal places.
Use $x_0 = 0$.

Steps	**Working**
1 Find $f'(x)$. Differentiate $y = x^3 + 2x + 1$.	
2 Calculate $f(x_0)$ and $f'(x_0)$ where $x_0 = 0$.	
3 Use Newton's method to find x_1. $x_{n+1} = x_n - \dfrac{f(x_n)}{f'(x_n)}$	
4 Calculate $f(x_1)$ and $f'(x_1)$ where $x_1 = -\dfrac{1}{2}$.	
5 Use Newton's method to find x_2. Note that $\begin{aligned} f(-0.4545) &= (-0.4545)^3 + 2(-0.4545) + 1 \\ &\approx -0.0029 \\ &\approx 0 \end{aligned}$	

FUNCTIONS AND GRAPHS

SB

p. 87

MATCHED EXAMPLE 1	The maximal domain

What is the maximal domain of $f(x) = \dfrac{x^2}{x^2 - 5x + 6}$?

Steps	Working
1 Factorise the denominator.	
2 Note values where $f(x)$ is undefined, i.e., where the expression in the denominator is 0.	
3 Write the answer.	

MATCHED EXAMPLE 2 | The range for a restricted domain

Find the range of the function $f: [2, 5) \to R, f(x) = x^2 - 6x + 5$.

Steps	Working
1 Write the quadratic coefficients.	
2 Find the zeros.	
3 Find the position of the minimum.	
4 Find the minimum.	
5 Find the values at the ends of the interval.	
6 Sketch the shape of the function.	
7 State the range.	

SB
p. 87

SB

Using CAS 1:
Range of a
function
p. 88

MATCHED EXAMPLE 3 The range for an unrestricted domain

Determine the range of the function $y = x^4 - 81$.

Steps	Working
1 Find the turning point.	
2 State the behaviour of the function.	
3 State the range.	

MATCHED EXAMPLE 4 | Finding if a function is one-to-one

Which of the following functions is one-to-one?

$f: [-2, 2) \rightarrow R, f(x) = x^3 + 4$

$g: R \rightarrow R, g(x) = (x + 2)(x - 3)(x + 1)$

$h: (-1, \infty) \rightarrow R, h(x) = \ln(x + 1)$

Steps	Working
1 Check f.	
2 Check g.	
3 Sketch h to determine if it is one-to-one.	
4 Check h.	
5 Write the final answer.	

MATCHED EXAMPLE 5 | Midpoint of two points

Find the midpoint of the line segment joining the points $(-2, 2)$ and $(-4, 6)$.

Steps	Working
1 Write the formula.	
2 Substitute values.	
3 Work out the answer.	

MATCHED EXAMPLE 6 | Distance between two points

Find the length of the line segment joining the points $(2, -2)$ to $(-1, 2)$.

Steps	Working
1 Write the formula.	
2 Substitute values.	
3 Simplify.	
4 Work out the answer.	

MATCHED EXAMPLE 7	Gradient of a line

Find the gradient of the straight line passing through $(1, 6)$ and $(3, 10)$.

Steps	Working
1 Write the formula.	
2 Substitute values.	
3 Simplify.	

MATCHED EXAMPLE 8 | Equation of a line through given points

Find the equation of the straight line passing through $(1, 4)$ and $(6, 6)$.

SB

p. 92

Steps	Working
1 Find the gradient.	
2 Write the point–gradient form.	
3 Substitute a point, say $(1, 4)$.	
It doesn't matter which point you substitute.	
4 Simplify.	
5 Write in general form.	

3

MATCHED EXAMPLE 9 | Equation of a line perpendicular to a given line

Find the equation of the straight line passing through $(-1, 4)$ that is perpendicular to $-3x + 4y + 12 = 0$.

Steps	Working
1 Find the gradient of the given line by writing it in gradient–intercept form.	
2 Write the gradient of the given line.	
3 Write the gradient of the new line.	
4 Write the point–gradient form.	
5 Substitute the values for the new line.	
6 Simplify and write in general form.	

SB

Using CAS 2:
Equation of a
perpendicular line
p. 93

MATCHED EXAMPLE 10	Sketching a quadratic that can be factorised

Sketch the graph of $f(x) = 4x - 2x^2 + 30$.

Steps	Working
1 Find the y-intercept.	
2 Find the zeros, if they exist.	
3 Find the turning point by completing the square.	
4 Sketch the graph.	

3

MATCHED EXAMPLE 11 | Sketching a quadratic without zeros

Sketch the graph of $f(x) = x^2 - 6x + 11$.

Steps	**Working**
1 Find the y-intercept.	
2 Find the zeros, if they exist.	
3 Find the turning point.	
4 Sketch the graph. $y = x^2$, translated 3 units to the right and 2 units up.	

MATCHED EXAMPLE 12 | Intersections of a straight line and quadratic

How many times does the line $y = 2x - 3$ intersect the parabola $f(x) = 5x - x^2 - 6$?

p. 99

Steps	Working
1 Make the equations equal each other.	
2 Simplify.	
3 The number of solutions to the quadratic equation will indicate the number of times the line intersects the parabola. Find the discriminant.	
4 State the result.	

Using CAS 3:
Intersections with
a straight line
p. 100

MATCHED EXAMPLE 13 | Sketching a quadratic using the quadratic formula

Sketch the graph of $y = x^2 - 10x + 7$.

Steps	**Working**
1 Find the y-intercept.	
2 Find the zeros using the quadratic formula, with $a = 1$, $b = -10$, $c = 7$.	
3 Find the turning point.	

The turning point is at $x = 5$, which is halfway between the zeros.

4 Sketch the graph.

Use the approximate values for *placement* of x-intercepts, but mark them with the exact values.

MATCHED EXAMPLE 14 | Graphs of power functions

Sketch the graph of $y = (x - 4)^6 - 7$ and state the implied domain and range.

Steps	Working
1 State the basic function.	
2 State the changes.	
3 Find the y-intercept.	
4 If possible, find any zeros.	
5 Sketch the graph.	
6 Use the graph to state the domain and range.	

p. 105

MATCHED EXAMPLE 15 | Graph of truncus

Sketch the graph of $y = 4 - \dfrac{1}{(x+1)^2}$ and state the maximal domain and range.

Steps	**Working**
1 State the basic function.	
2 State the changes.	
3 Find the y-intercept.	
4 State the asymptotes.	
5 Find the zeros.	
6 Sketch the graph, including the translated vertices.	
7 Use the graph to state the domain and range.	

MATCHED EXAMPLE 16 | Graph of hyperbola

a Show that $y = \dfrac{2x+1}{x+1}$ can be written in the form $y = \dfrac{a}{x-1} + b$, where a and b are constants.

b Hence, sketch the graph of $y = \dfrac{2x+1}{x+1}$ and state the natural domain and range.

Steps	Working
a Change to given form by making the numerator a multiple of the denominator and adjusting.	
b 1 State the basic function.	
2 State the changes.	
3 Find the y-intercept.	
4 Find any zeros.	
5 State the asymptotes.	
6 Sketch the graph, including the translated vertices.	
7 Use the graph to state the domain and range.	

3

MATCHED EXAMPLE 17 Graph of square root function

Sketch the graph of $y: D \to R, \ y = \sqrt{x-2} - 2$, where D is the maximal domain of f.

State the maximal domain and range.

Steps	Working
1 State the basic function.	
2 State the changes.	
3 Use the changes to find the new position of the starting point.	
4 Find the y-intercept.	
5 Find the zero.	
6 Sketch the graph.	
7 Use the graph to state the domain and range.	

MATCHED EXAMPLE 18	Graph of cube root function

SB

p. 107

Sketch the graph of $y: R \to R$, $y = \sqrt[3]{x+1} - 3$ and state the natural domain and range.

Steps	Working
1 State the basic function.	
2 State the changes.	
3 State the point of inflection.	
4 Find the y-intercept.	
5 Find the zero.	
6 Sketch the graph.	
7 Use the graph to state the domain and range.	

3

SB

p. 110

MATCHED EXAMPLE 19 | Existence of an inverse

Does an inverse function exist for $f: R \to R,\ f(x) = \dfrac{1}{2}x^4 - 1$?

Steps	Working
1 State the condition necessary.	
2 Turn this around for the original function.	
3 State a counterexample.	
4 State the result.	

Find the inverse function for $f: R \to R, \ f(x) = 2x^3 - 5$.

SB
p. 110

Steps	Working
1 Swap x and y.	
2 Solve for y.	
3 Write the result with the domain and range.	

SB

Using CAS 4:
Finding the inverse
of a function
p. 111

p. 111

MATCHED EXAMPLE 21 | Drawing an inverse from the graph

The graph of the function $y = f(x)$ is sketched below. Sketch the graph of the inverse function.

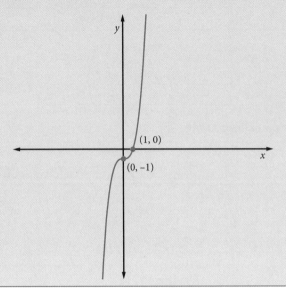

Steps	Working
Reflect the graph in the line $y = x$.	

MATCHED EXAMPLE 22	Changing an exponential to a logarithm	

If $10^{x-3} = 5y - 2$, find an expression for x.

SB

p. 116

Steps	Working
1 Change to a logarithm.	
2 Make x the subject.	

MATCHED EXAMPLE 23 | Changing a natural logarithm to an exponential

Find x, given that $\log_e (5x+7) = 7 - y^3$.

Steps	Working
1 Change to an exponential.	
2 Make x the subject.	

MATCHED EXAMPLE 24	Simplifying a logarithmic expression	

Simplify $\log_5(2)\log_3(5)\log_6(6)$.

p. 117

Steps	**Working**
1 Change all to base 3 using the change of base theorem.	
2 Simplify.	

MATCHED EXAMPLE 25 | Logarithmic graphs

Part of the graph of $f: R \to R$, $f(x) = 5^{2x+1} - 2$ is shown below.

Sketch the corresponding part of the inverse function f^{-1}.

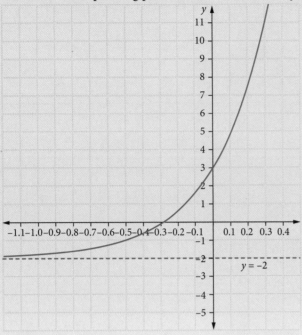

Steps	**Working**

1 Identify points on the original graph.

2 State the inverse points.

3 Plot the inverse points on the new graph and join with a smooth curve. Label the new function as $f^{-1}(x)$

The inverse of the asymptote $y = -2$, is the asymptote $x = -2$.

9780170464062

MATCHED EXAMPLE 26	Degrees and radians

Change 70° to radians.

p. 121

Steps	Working
Use the basic conversion.	

MATCHED EXAMPLE 27 | Exact value

Find the exact value of $\cos\left(\dfrac{5\pi}{4}\right)$.

Steps	**Working**
1 Use ASTC to establish the sign.	
2 Write the angle using an acute angle.	
3 Give the exact value.	

MATCHED EXAMPLE 28 | Finding a second trigonometric ratio

$\cos(x) = \dfrac{4}{5}$. Find the exact value of $\cot(x)$, given that $\dfrac{3\pi}{2} \leq x \leq 2\pi$.

Steps	Working
1 Write the identity to find $\sin(x)$.	
2 Solve.	
3 Find $\cot(x)$.	
4 Use the quadrant information.	

Alternative method

1 Draw a right-angled triangle using $\cos(\theta) = \dfrac{4}{5}$.
Make the base 4 and the hypotenuse 5
and θ the acute angle corresponding to x.

2 Find the value of the third side.

3 Find $\cot(\theta)$ using the triangle.

4 Now, use the fact that x is in the fourth quadrant.

SB

p. 125

MATCHED EXAMPLE 29 | Circular function graph on a limited domain

Sketch the graph of $y = 5\sin(3x) + 2$ for $-\pi < x \le \pi$.

Steps	Working
1 Find the period.	
2 State the amplitude.	
3 State the mean value.	
4 Sketch the graph. Make sure you show that $(-\pi, 2)$ is not included but $(\pi, 2)$ is included.	

MATCHED EXAMPLE 30 | Period, amplitude and mean value of a circular function

A function is given by $f: R \to R, f(x) = 3\cos\left(\dfrac{2\pi x}{5}\right) + 1$.

State the period, amplitude and mean value of the function.

Steps	Working
1 Find the period T.	
2 State the amplitude.	
3 State the mean value.	

MATCHED EXAMPLE 31 | Writing an equation for a circular function

The diagram below shows part of a circular function. State a possible equation for the graph.

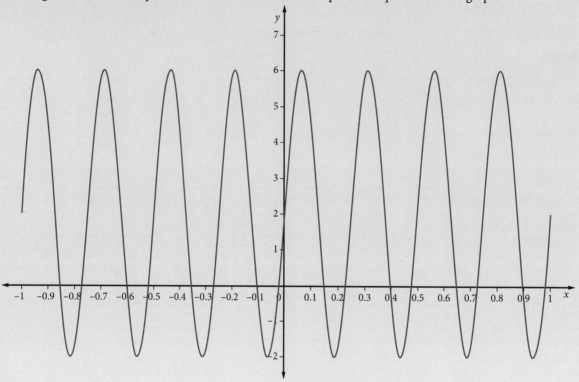

Steps	Working
1 Find the period T.	
2 Find the amplitude.	
3 Find the mean value.	
4 Identify the circular function.	
5 Write the general form and find the value of n.	
6 Substitute values and simplify.	
7 Write the answer.	

DIFFERENTIATION

MATCHED EXAMPLE 1	Average rate of change

p. 144

Find the average rate of change between $x = 3$ and $x = 10$ for the function with the rule $y = 2x^2 + 3$.

Steps	Working
1 Work out the required y values.	
2 Write the coordinates.	
3 Calculate the gradient.	
4 State the average rate of change.	

MATCHED EXAMPLE 2 | Differentiation by first principles

Use differentiation by first principles to find the instantaneous rate of change of the function $f(x) = x^4$ at $x = 3$.

Steps	Working
1 Write the function.	
2 Find an expression for $f(x + h)$. Expand and simplify.	
3 Find an expression for $\dfrac{f(x+h)-f(x)}{h}$ and simplify.	
4 Find $\displaystyle\lim_{h \to 0} \dfrac{f(x+h)-f(x)}{h}$.	
5 Evaluate $f'(x)$ using the given value.	

Explain why each function is not differentiable at the given point.

a $f(x) = \dfrac{1}{\tan x}$ at $x = 0$

b $f(x) = \dfrac{x^3 - 27}{x - 3}$ at $x = 3$

c $f(x) = |x - 2|$ at $x = 2$

Steps	Working
a 1 Decide if the function is continuous.	
2 Decide if the limit is differentiable.	
b 1 Simplify the function.	
2 Decide if the function is continuous.	
3 Decide if the limit is differentiable.	
c 1 Decide if the function is continuous.	
2 Decide if the limit is differentiable.	

MATCHED EXAMPLE 4 | The derivative of ax^n

Differentiate each function.

a $f(x) = 2x^3 + 3\sqrt{5x^{\frac{11}{2}}}$

b $f(x) = 2x^9$

c $f(x) = \dfrac{5}{7x^{\frac{5}{4}}}$

Steps	Working
a **1** Write in the form $f(x) = ax^n$.	
2 Differentiate using $f'(x) = a \times nx^{n-1}$.	
3 Simplify.	
b **1** Write in the form $f(x) = ax^n$.	
2 Differentiate using $f'(x) = a \times nx^{n-1}$.	
3 Simplify.	
c **1** Write in the form $f(x) = ax^n$.	
2 Differentiate using $f'(x) = a \times nx^{n-1}$.	
3 Express with positive indices.	

For each function, $f(x)$, calculate $f'(x)$ using the given value of x.

a $f(x) = 7x^5 + 3x^3 + 3$, $f'(1)$

b $f(x) = \dfrac{36 - x^2}{x + 6}$, $f'(2)$

c $f(x) = \dfrac{5x^{\frac{8}{3}} + 3x^{\frac{14}{3}}}{x^{\frac{5}{3}}}$, $f'\left(\dfrac{2}{3}\right)$

Steps	**Working**
a 1 Write $f(x)$ in polynomial form.	
2 Differentiate each term.	
3 Evaluate $f'(x)$.	
b 1 Write $f(x)$ in polynomial form.	
2 Differentiate each term.	
3 Evaluate $f'(x)$.	
c 1 Write $f(x)$ in polynomial form.	
2 Differentiate each term.	
3 Evaluate $f'\left(\dfrac{2}{3}\right)$.	

SB p. 150

4

Using CAS 1:
The derivative
p. 150

Using CAS 2:
Instantaneous rate
of change
p. 151

p. 155

MATCHED EXAMPLE 6 The product rule

Use the product rule to differentiate $f(x) = (3x^5 + x)(2x^2 - 11)$.

Steps	Working
1 Identify u and v.	
2 Differentiate to obtain u' and v'.	
3 Write down the expression for $uv' + vu'$.	
4 Expand and simplify.	
5 Write the answer.	

9780170464062

MATCHED EXAMPLE 7 | The product rule with substitution

For the function $f(x) = (x^2 + 2x - 5)(3x^3 + 2x^2 - 5x - 10)$, find $f'(-1)$.

Steps	Working
1 Identify u and v.	
2 Differentiate to obtain u' and v'.	
3 Give the expression for $u'v + uv'$.	
4 Substitute the value and simplify.	

MATCHED EXAMPLE 8 | The quotient rule

Find the derivative of $\dfrac{5x+2}{7x-1}$.

Steps	**Working**
1 Identify u and v.	
2 Differentiate to obtain u' and v'.	
3 Write down the expression for $\dfrac{vu'-uv'}{v^2}$.	
4 Expand and simplify.	
5 Write the answer.	

MATCHED EXAMPLE 9 | The chain rule

Differentiate each function with respect to x using the chain rule.

a $\dfrac{1}{(7x+5)^3}$

b $\sqrt{3x-1}$

Steps	**Working**
a 1 Write $\dfrac{1}{(7x+5)^3}$ as a function of a function in index form.	
2 Write the chain rule.	
3 Substitute the derivatives.	
4 Substitute for u.	
5 Write the answer.	
b 1 Write $\sqrt{3x-1}$ as a function of a function.	
2 Write the chain rule.	
3 Substitute the derivatives.	
4 Substitute for u.	
5 Write the answer.	

MATCHED EXAMPLE 10 | Combining the chain and product rules

Differentiate $3x^4(2x + 5)^2$.

Steps	Working
1 Write as a product.	
2 Find the derivative of u.	
3 Write v as a function of a function.	
4 Write the chain rule.	
5 Substitute the derivatives.	
6 Substitute for q.	
7 Write the product rule.	
8 Substitute the functions.	
9 Take out the common factor.	
10 Simplify and write the answer.	

FUNCTIONS AND TRANSFORMATIONS

MATCHED EXAMPLE 1 | Dilation from the *y*-axis

SB

p. 179

Draw the graph of $y = \left(\dfrac{x}{3} + 1\right)^3$ on the same set of axes as $y = (x+1)^3$ and describe the transformation from x to $\dfrac{x}{3}$.

Steps	Working
1 Both graphs are of cubic functions. Compile a table of values from, say, 0 to 8.	
2 Sketch both graphs on the same axes, using a smooth line to connect the points.	
3 Describe the transformation.	

SB

Using CAS 1:
Dilation from the
x-axis
p. 176

SB

Using CAS 2:
Dilation from the
y-axis
p. 177

p. 180

MATCHED EXAMPLE 2 | Reflections in the axes

The graph of $y = f(x)$ is shown at right.
Sketch the graphs of $y = f(-x)$ and $y = -f(x)$
on separate sets of axes.

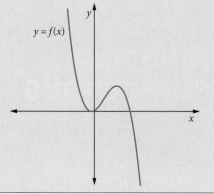

Steps	Working
1 $y = f(-x)$ is reflected in the y-axis.	
A negative x reflects the graph horizontally	
2 $y = -f(x)$ is reflected in the x-axis.	
A negative $f(x)$ reflects the graph vertically.	

MATCHED EXAMPLE 3 | Translations

The graph of $y = f(x)$ is shown.

Sketch the graphs of $y = f(x + 2)$ and $y = f(x) - 2$ on separate sets of axes.

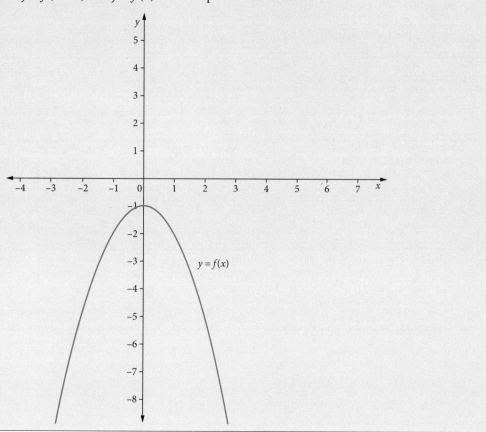

Steps	Working

1 $y = f(x + 2)$ is translated 2 units in the negative direction of the x-axis (to the left) from $y = f(x)$.

> A value added to or subtracted from x translates the graph horizontally.

2 $y = f(x) - 2$ is translated 2 units in the
negative direction of the y-axis (down)
from $y = f(x)$.

A value added to or subtracted
from $f(x)$ translates the graph
vertically.

9780170464062

MATCHED EXAMPLE 4 | Multiple transformations

A function $f(x)$ is translated 6 units in the negative x direction, reflected in the x-axis and dilated from the y-axis by a factor of $\dfrac{1}{2}$, in that order. The point $P(0, 2)$ is on the graph of $f(x)$. What are the coordinates of the point Q, to which it is transformed?

p. 182

Steps	Working
1 Translate 6 units in the negative x direction.	
2 Reflect in the x-axis.	
3 Dilate from the y-axis by a factor of $\dfrac{1}{2}$.	
4 Write the answer.	

p. 182

MATCHED EXAMPLE 5 | Inverse transformations

The function $f(x) = \sqrt{x}$ is transformed to $g(x) = \sqrt{2-3x} - 4$.

Describe the inverse transformation that transforms $g(x)$ back to $f(x)$.

Steps	Working
1 Show the transformation as a series of steps.	
2 Describe the steps.	
3 Reverse the steps.	
4 Write the description of the inverse transformation.	

9780170464062

MATCHED EXAMPLE 6 | Combined transformations

The graph of the function $y: R-\{0\} \to R, \ y = \dfrac{1}{x}$ is translated to the right by 4 units, translated 1 unit in the negative y direction, reflected across the x-axis, dilated from the x-axis by a factor of 3 and dilated parallel to the parallel to the x-axis by a factor of $\dfrac{1}{2}$. State the rule of $f(x)$, sketch the function and give its maximal domain and range.

Steps	Working
1 Translate to the right and 1 unit down by replacing x by $(x-4)$ and f by $f-1$.	
2 Reflect parallel to the y-axis by multiplying -1 by the (new) f term	
3 Dilate parallel to the y-axis by multiplying the (new) f term by factor of 3: $3 \times f$.	
4 Dilate from the y-axis by multiplying the (new) x term by 2.	
5 Sketch the function. The asymptote $x = 0$ moves 4 units right to $x = 4$ and is then dilated vertically by a factor of $\dfrac{1}{2}$ to $x = 2$. The x-intercept is given by $3 - \dfrac{3}{2x-4} = 0$, which gives $x = \dfrac{5}{2}$. The y-intercept is given by $3 - \dfrac{3}{2(0)-4} = y$, which gives $y = \dfrac{15}{4}$.	
6 State the domain and range.	

p. 189

MATCHED EXAMPLE 7 | Describing combined transformations

Specify the transformations performed on $f(x)$ to give $g(x)$.

a $f(x) = x^2 \rightarrow g(x) = 5 - \dfrac{(x-2)^2}{3}$

b $f(x) = 4 - \sqrt{-x+3} \rightarrow g(x) = \sqrt{x}$

c $f(x) = \dfrac{1}{x} \rightarrow g(x) = \dfrac{3}{1-0.5x} - 4$

Steps	Working
a **1** Write $g(x)$ in the standard form.	
2 Write in the form $af(n(x+b)) + c$.	
3 Write the transformations.	
b **1** Write $g(x)$ as a transformation of $f(x)$. Do the 'outside' transformations first.	
2 Now do the 'inside' transformations. Replace x by $n(x + b)$ in the expression of f to cancel back to $g(x)$.	
3 Specify the transformations in order.	
c **1** Write $g(x)$ in the form $af(n(x + b)) + c$.	
2 Specify the transformations in order.	

MATCHED EXAMPLE 8 | The rule of a function after a combined transformation

State a possible rule for the function $f(x)$ whose graph is shown below, and specify the transformations performed on the basic function.

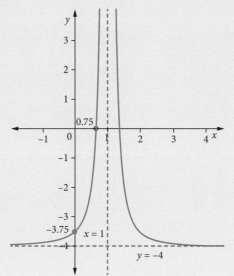

Steps	Working
1 The graph is a truncus, the reciprocal of the square function.	
2 The y-intercept is 1 unit to the left and 0.25 units up from the intersection of the asymptotes.	
3 Perform a dilation of $\dfrac{1}{x^2}$ to get $(-1, 0.25)$.	
4 The intersection of the asymptotes $(0, 0)$ is translated 1 unit to the right and 4 units down.	
5 Simplify and write as a function.	
7 Check the known points $(0.75, 0)$ and $(0, -3.75)$.	
8 Write $f(x)$ as a transformation of $b(x) = \dfrac{1}{x^2}$.	
9 Specify the transformations.	

SB

Using CAS 3:
Transformations
of graphs
p. 190

MATCHED EXAMPLE 9 | Finding rules for transformations of graphs

The graph of a function $f(x)$ is shown below. Its rule is given by

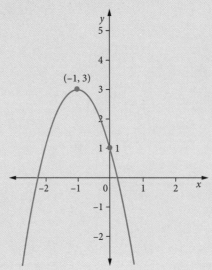

A $f(x)=-(2x+1)^2+3$ **B** $f(x)=-\left(\dfrac{x}{2}+1\right)^2+2$ **C** $f(x)=-2(x+1)^2-3$

D $f(x)=-2(x-1)^2+3$ **E** $f(x)=-2(x+1)^2+3$

Steps	Working
1 Check $f(0)$ and $f(-1)$ for **A**.	
2 Check $f(0)$ and $f(-1)$ for **B**.	
3 Check $f(0)$ and $f(-1)$ for **C**.	
4 Check $f(0)$ and $f(-1)$ for **D**.	
5 Check $f(0)$ and $f(-1)$ for **E**.	
6 Write the answer.	

MATCHED EXAMPLE 10 | Writing matrices for transformations

Find the transformation matrices for each sequence of transformations.

a $x^2 \rightarrow 4(-x-1)^2 + 3$

b $-\sin(3x+2) - 5 \rightarrow \sin(x)$

p. 196

Steps	Working
a 1 Write the transformations in order, $x^2 \rightarrow 4x^2 \rightarrow 4(-x)^2 \rightarrow 4(-(x+1))^2 + 3 = 4(-x-1)^2 + 3$	
2 Write the transformation matrices in order.	
3 Simplify.	
b 1 Write the transformations in order.	

Do the opposite of the transformation:
$\sin(x) \rightarrow -\sin(3x+2) - 5$,
$\sin(x) \rightarrow -\sin(x) \rightarrow -\sin(x+2) - 5 \rightarrow$
$\rightarrow (\sin(3x+2) + 5) = -\sin(3x+2) - 5$

2 Write the transformation matrices in order.	
3 Simplify.	

MATCHED EXAMPLE 11 | Applying transformation matrices with translation first

The transformation $T: R^2 \to R^2$, $T\left(\begin{bmatrix} x \\ y \end{bmatrix}\right) = \begin{bmatrix} \pi & 0 \\ 0 & -\dfrac{1}{2} \end{bmatrix}\left(\begin{bmatrix} x \\ y \end{bmatrix} + \begin{bmatrix} -\dfrac{1}{4} \\ 2 \end{bmatrix}\right)$ is applied to the curve with

equation $y = \cos(x)$. What is the image of the curve under the transformation?

Steps	Working
1 Apply the transformation.	
2 Do the brackets first.	
3 Now do the multiplication.	
4 Separate x' and y'.	
5 Make x and y the subjects.	
6 Substitute in the equation of the curve $y = \cos(x)$.	
7 Make y' the subject.	
8 Simplify.	
9 Write the answer.	

MATCHED EXAMPLE 12 | Applying transformation matrices with translation last

$T: R^2 \to R^2$, $T\left(\begin{bmatrix} x \\ y \end{bmatrix}\right) = \begin{bmatrix} -\dfrac{1}{2} & 0 \\ 0 & 1 \end{bmatrix} \begin{bmatrix} x \\ y \end{bmatrix} + \begin{bmatrix} 3 \\ 2 \end{bmatrix}$ maps a function f to the curve with equation

$y = 2(x-1)^2 + 3x$. What is the equation of f?

Steps	Working
1 Apply the transformation.	
2 Do the multiplication first.	
3 Now do the addition.	
4 Separate x' and y'.	
5 Substitute into the image equation $y' = 2(x'-1)^2 + 3x'$.	
6 Simplify the brackets.	
7 Make y the subject.	
8 Write the answer.	

SB

Using CAS 4:
Matrices and
transformations
p. 198

MATCHED EXAMPLE 13 | Adding functions

The functions $f: [-5, 5] \rightarrow R, f(x) = 5(2x-1)^4 + 4$ and $g: (-\infty, 0) \rightarrow R, g(x) = -96\left(x - \dfrac{1}{2}\right)^4 - 3$ are added together.

a What is the rule for the combined function $f(x) + g(x)$, and what is its maximal domain?

b What is the maximal domain of the inverse of the combined function from part **a**?

Steps	Working
a 1 Add the functions.	
2 Find the intersection of the domains.	
b 1 For the range of $f(x) + g(x)$, first check its turning point.	
2 Find the combined range.	
3 The domain of the inverse function is the range of the original function.	

Find a pair of functions whose difference could give $y = \dfrac{1}{2}\log_3\left[16x^2(x+2)^{-2}\right]$.

SB

p. 204

Steps	Working
1 Rewrite the negative exponent in the function as a positive exponent. Use the logarithm law: $\log_a\left(\dfrac{x}{y}\right) = \log_a(x) - \log_a(y)$	
2 Apply to the function.	
3 Simplify each term.	
4 Write the answer.	

5

MATCHED EXAMPLE 15 | Combining graphs

The graphs of $y = f(x)$ and $y = g(x)$ are shown below at the same scale over the same domain. Sketch the graphs of

a $f(x) + g(x)$ **b** $f(x) \times g(x)$

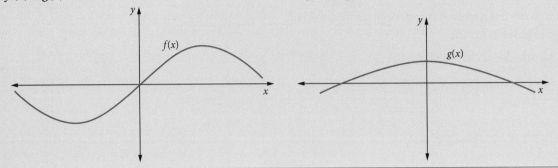

Steps	Working
a 1 On the left, $f(x)$ is negative and $g(x)$ is negative.	
2 $g(x)$ changes to positive while $f(x)$ is still negative and increasing.	
3 $f(x)$ and $g(x)$ both increase a bit past the negative x-axis.	
4 $g(x)$ is decreasing and positive and $f(x)$ is increasing and positive after $x = 0$.	
5 $f(x)$ increases rapidly and its magnitude quickly becomes greater than $g(x)$.	
6 Sketch the graph of $f(x) + g(x)$.	
b 1 The zeros of $f \times g$ will be the zeros of f and of g combined. Mark the positions of the zeros.	
2 Use the signs of f and g to find the signs of $f \times g$ zeros.	

3 Use the zeros and signs to sketch the probable shape of the graph.

MATCHED EXAMPLE 16	Composite function values

$f(-1) = 0, f(0) = 2, f(2) = -2, f(3) = 3, f(-2) = -1, g(-1) = 3, g(0) = 2, g(2) = -2, g(3) = -1, g(-2) = 0$

What is the value of

a $f(g(-2))$?　　　　**b** $g(f(-2))$?　　　　**c** $g \circ f(0)$?　　　　**d** $f \circ g(0)$?

Steps	Working
a **1** Look up the value of $g(-2)$.	
2 Now find the value of $f(0)$.	
3 Write the answer.	
b **1** Look up the value of $f(-2)$.	
2 Now find the value of $g(-1)$.	
c **1** Write the meaning of $g \circ f(0)$.	
2 Look up the value of $f(0)$.	
3 Now find the value of $g(2)$.	
d **1** Look up the value of $g(0)$.	
2 Now look up the value of $f(2)$.	

SB

p. 209

$f(x) = x^2 + 2x + 4$ and $g(x) = \dfrac{3}{x}$. Let $h = f \circ g$.

a Find the value of $f \circ g(1.5)$.

b Find the rule for h.

c Find the maximal domain and range of h.

d Find the value of $h(0.25)$.

Steps	Working
a 1 Find $g(1.5)$.	
2 Find $f(2)$.	
3 Write the answer.	
b 1 Write the rule for $h(x)$.	
2 Replace $g(x)$ by its rule.	
3 Put $\dfrac{3}{x}$ into the rule for f.	
4 Simplify if possible.	
c 1 Consider the domain of $f(x)$.	
2 Consider the range of $g(x)$. Check that the range of $g(x)$ is not larger than the domain of $f(x)$.	
3 The values of g are the inputs for f, so the values of g must be restricted.	
4 Write the domain.	
5 Write the range.	
d 1 Calculate $h(0.25)$.	
2 Write the answer.	

MATCHED EXAMPLE 18 | Composite graphs

The graphs of $f(x)$ and $g(x)$ are shown below. Sketch a possible graph for $f(g(x))$.

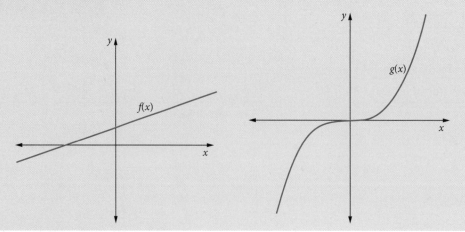

Method 1: Use general values of graphs

Steps	**Working**
1 Consider x negative and of large magnitude.	
2 Consider x negative and $x \to 0$ ($x \uparrow 0$).	
3 Consider what happens to $f(X)$ as $g(x)$ approaches its minimum.	
4 Consider what happens as x increases from the minimum value ($x \to \infty$).	
5 Consider what happens to $f(X)$ as $g(x)$ increases and approaches ∞.	
6 Sketch the graph, so that $f(X)$ has a negative x-intercept and increases to reach the positive y-intercept.	

Method 2: Use possible rules for the functions	
Steps	**Working**
1 State the nature of $f(x)$.	
2 State a possible equation for $f(x)$.	
3 State the nature of $g(x)$.	
4 State a possible equation for $g(x)$.	
5 Work out a possible $f(g(x))$.	
6 Sketch a possible graph for $f(g(x))$.	

SB

Using CAS 5:
Composite
functions
p. 211

SB

p. 212

MATCHED EXAMPLE 19 | Finding rules for transformations of graphs

$f(x) = x^2 + x - 1$ and $g(x) = x + 2$

a Find the rule for $f \circ g$ and state its domain and range.

b Find the rule for $g \circ f$ and state its domain and range.

c The transformation $T: \begin{bmatrix} x \\ y \end{bmatrix} \rightarrow \begin{bmatrix} a & 0 \\ 0 & b \end{bmatrix} \begin{bmatrix} x \\ y \end{bmatrix} + \begin{bmatrix} c \\ d \end{bmatrix}$ maps $f \circ g(x)$ to $g \circ f(x)$. Find the values of a, b, c and d.

Steps	Working
a 1 Use substitution to find the rule.	
2 Simplify.	
3 Complete the square to find the range.	
4 State the answer.	
b 1 Use substitution to find the rule.	
2 Complete the square to find the range.	
3 State the answer.	
c 1 Use $y = f \circ g(x)$ and $y' = g \circ f(x')$	
2 Separate transformations parallel to each axis.	
3 Simplify the transformation of y.	
4 Simplify the transformation of x.	
5 Write the values of a, b, c, d.	

MATCHED EXAMPLE 20 | Graphing a piecewise function

Sketch the graph of the function $f(x) = \begin{cases} x^2 - 2 & \text{for } -5 \le x < -1 \\ x^3 + 2 & \text{for } -1 \le x < 3 \\ 2^x & \text{for } x \ge 3 \end{cases}$ from $x = -5$ to $x = 5$.

Steps	Working
1 Find the extreme values for each section.	
2 Choose suitable scales to sketch the graph. Make sure you show the endpoints as included or excluded according to the function definition.	

     Chapter 5 | Functions and transformations   **107**

MATCHED EXAMPLE 21	Using a piecewise function

On a shopping website, the prices of dinner spoons vary according to the number of spoons ordered. For orders of 12 or fewer spoons, the price is $2.56 per spoon plus the shipping charge of $6. For orders of more than 12 spoons, the price is $1.92 per spoon and free shipping. State a price function for ordering dinner spoons. Find the price for ordering 7 spoons and 24 spoons.

Steps	Working
1 Write the definition in two parts.	
2 Use the function for $x = 7$.	
3 Use the function for $x = 24$.	

APPLICATIONS OF DIFFERENTIATION

MATCHED EXAMPLE 1 | Positive and negative gradients

p. 238

State the x value(s) where the gradient of the graph of $f(x) = x(x-1)^2(x+2)^2$

a is positive.

b is negative.

Steps	Working
a 1 Find the derivative of the expression $x(x-1)^2(x+2)^2$ and equate to zero.	

TI-Nspire   **ClassPad**

2 Sketch the graph of $f(x) = x(x-1)^2(x+2)^2$ to identify four turning points.

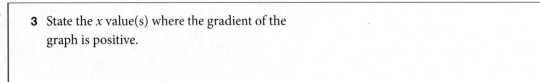

SB

Using CAS 1:
Differentiation
p. 236

3 State the x value(s) where the gradient of the graph is positive.

b State the x value(s) where the gradient of the graph is negative.

MATCHED EXAMPLE 2 | Strictly increasing and decreasing

State the interval over which the function with the rule $f(x) = 2(x + 1)^2(\frac{x}{2} - 1)$ is

a strictly increasing.

b strictly decreasing.

Steps	Working
1 Find the derivative of the expression $2(x + 1)^2(\frac{x}{2} - 1)$ and equate to zero. **TI-Nspire**　　　　　　　**ClassPad**	
2 Sketch the graph of $f(x) = 2(x + 1)^2(\frac{x}{2} - 1)$ to identify two turning points.	
3 a State the x value(s) for which the graph of $f(x)$ is **strictly increasing**. 　　**b** State the x value(s) for which the graph of $f(x)$ is **strictly decreasing**.	

p. 242

MATCHED EXAMPLE 3 | Turning points

Find the coordinates and nature of the turning points on the graph of $f(x) = \dfrac{5}{2}x^3 + 3x^2$.

Steps	Working
1 Sketch the graph of $f(x) = \dfrac{5}{2}x^3 + 3x^2$.	
2 Find $f'(x)$ and solve for zero.	
3 Check whether the sign of the gradient changes on either side to identify the nature of the stationary points.	
4 Find the coordinates and nature of the turning points.	

Using CAS 2:
Stationary points
p. 243

MATCHED EXAMPLE 4 | Stationary points of inflection

Find the coordinates of any stationary points of inflection on the graph $f(x) = (x - 2)^3(x + 2)$.

SB

p. 245

Steps	**Working**
1 Sketch the graph of $f(x) = (x - 2)^3(x + 2)$.	
2 Find $f'(x)$ and solve for zero.	
TI-Nspire	**ClassPad**
3 Check to see if the sign of the gradient changes or stays the same on either side to identify the nature of the stationary points.	
4 Find the coordinates of the stationary point of inflection.	

SB

p. 249

MATCHED EXAMPLE 5	Global maximum and minimum

Find the coordinates of the global maximum for the function $f: [-3, 2] \to R, f(x) = 2(\frac{x}{2} + 1)^2(2 - x)$.

Steps	Working

1 Sketch the graph of
$f: [-3, 2] \to R, f(x) = 2(\frac{x}{2} + 1)^2(2 - x)$.

2 Find $f'(x)$ and solve for zero.

TI-Nspire

ClassPad

3 Identify the point that is a local maximum turning point.

4 Test the endpoints of the interval $[-3, 2]$.

5 Find the coordinates of the global maximum.

9780170464062

MATCHED EXAMPLE 6 | Graphing a function

Sketch the graph of $f: [-2, 2] \to R$, $f(x) = -3(x + 2)^2(x - 1)$, labelling the key features. State the range of f.

Steps	Working
1 Consider the general shape.	
2 Find the y-intercept.	
3 Find the x-intercept(s).	
4 Solve $f'(x) = 0$ for stationary points.	
5 Calculate the y values of the stationary points.	
6 Calculate the y values of the endpoints.	
7 Sketch the graph for the domain.	
8 State the range.	

MATCHED EXAMPLE 7 | Tangent to a graph

a Find where the tangent to the graph of $y = x^3 + x^2 + 2$ at the point $(1, 4)$ meets the graph again.

b Sketch the graph of both the curve and its tangent, showing the intersection points.

Steps	Working
a 1 For the curve $y = x^3 + x^2 + 2$, find $\dfrac{dy}{dx}$ at $x = 1$.	
2 Use $y - y_1 = m(x - x_1)$ to find the equation of the tangent at point $(1, 4)$.	
3 Find the points of intersection of the curve and the tangent. The question already says one of the points is $(1, 4)$.	
4 Find the coordinates of the points at which the tangent meets the graph again.	
b Sketch the graph.	

SB

Using CAS 3:
Tangent to a graph
p. 254

SB

p. 258

A circular cone has a height of x cm, where the radius is 4 cm minus the height. Find the maximum volume, in cm^3, correct to one decimal place, for this cone.

Steps	Working
1 Sketch a diagram.	
2 Set up an equation to describe the volume of the cone.	
3 Solve $V'(x) = 0$ for local maximum and minimum values.	
4 Identify the domain for this problem.	

5 Consider the graph shape to decide which value of x gives the maximum value and whether it lies in the domain $(0, 4)$.

Using CAS 4:
Maximum
and minimum
problems
p. 259

6 Substitute this value of x into $V(x)$ to find the maximum volume of the cone.

MATCHED EXAMPLE 9 | Graphing derivative functions

The graph of $y = f(x)$ is shown below. Sketch the graph of $y = f'(x)$.

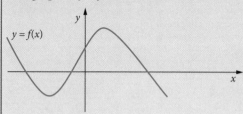

Steps	**Working**
• The function is decreasing at first, so the derivative graph is negative (below the x-axis).	
• There is a stationary point just before the y-axis, so the derivative graph is zero (x-intercept).	
• Then the function is increasing, so the derivative graph is positive (above the x-axis).	
• At the second stationary point, the derivative graph is zero again.	
• Then the function is decreasing again, so the derivative graph is negative again.	
Graph the derivative function so that the points match the relevant points on the original function.	

SB

p. 265

MATCHED EXAMPLE 10 Graphing derivative functions with discontinuities

Given the graph below, sketch the graph of its derivative function.

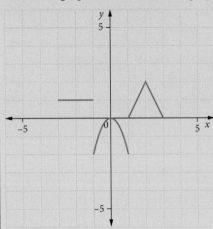

Steps	Working
1 Note the significant features of the given graph.	
2 Sketch the derivative graph. Note the discontinuous points.	

CHAPTER (7)

INTEGRATION

MATCHED EXAMPLE 1	Finding the anti-derivative

Find the anti-derivative of the function $y = 4x^3 - x - 7$.

SB

p. 281

Steps	**Working**
1 Write the function as a derivative.	
2 Integrate each term using the formula $\int ax^n dx = \dfrac{ax^{n+1}}{n+1} + c.$	
3 Simplify the expression.	

MATCHED EXAMPLE 2 Integral of $(ax + b)^n$

Find $\int (7x+3)^5 \, dx$.

Steps	Working
1 Integrate using $$\int (ax+b)^n \, dx = \frac{(ax+b)^{n+1}}{a(n+1)} + c$$	
2 Simplify.	

We can skip Step 1 and go straight to the answer.

MATCHED EXAMPLE 3 | Finding y given $\dfrac{dy}{dx}$

Find y if $\dfrac{dy}{dx} = 4x^3 + 2x - 1$ and $x = 3$ when $y = 0$.

Steps	Working
1 Write the expression as a derivative.	
2 Integrate each term and simplify.	
3 Find the value of c by substituting $x = 3$, $y = 0$.	
4 State the answer including the value of c.	

SB

Using CAS 1:
The anti-derivative
p. 284

p. 285

MATCHED EXAMPLE 4 | Straight-line motion

The acceleration of a vehicle travelling in a straight line is given by $a(t) = 4t + 2$.

a Find an expression for the velocity $v(t)$ if $v = 0$ when $t = 1$.

b Find an expression for the displacement $x(t)$ if the vehicle started at the origin.

Steps	Working
a 1 State the rule for acceleration.	
2 Use velocity $= v = \int a(t)\,dt$.	
3 Find c using $v = 0$ when $t = 1$.	
4 State the velocity function.	
b 1 Use displacement $= x = \int v(t)\,dt$.	
2 Find d using $x = 0$ when $t = 0$.	
3 State the displacement function.	

MATCHED EXAMPLE 5 | Graphing *f*(*x*) given *f*′(*x*)

The graph of $y = f'(x)$ is shown below. Sketch a possible graph of $y = f(x)$.

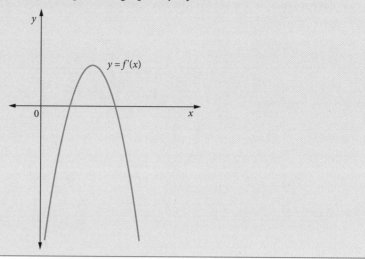

Steps	Working

1 Look at the *sign* of the derivative graph $y = f'(x)$.

- It starts off negative (below the *x*-axis), which means that the function is decreasing.

- It becomes zero, then positive, so the function reaches a minimum point, then is increasing.

- It becomes zero again, then negative, so the function reaches a maximum point, then is decreasing again.

2 Graph $y = f(x)$ so that the points match the relevant points on the derivative function.

MATCHED EXAMPLE 6 | Approximation using left rectangles

Find an approximation to the area under the curve $y = \frac{1}{2}(x-1)^2$ between $x = 3$ and $x = 5$ using the sum of two rectangles shown below.

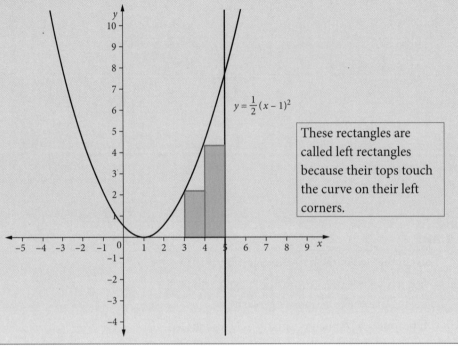

$y = \frac{1}{2}(x-1)^2$

These rectangles are called left rectangles because their tops touch the curve on their left corners.

Steps	Working
1 Find the height of each rectangle.	
2 Find the area of each rectangle.	
3 Add the areas.	

MATCHED EXAMPLE 7 | Approximation using rectangles

SB

p. 292

Find an approximation to the area under the curve $y = 7 - \dfrac{1}{4}x^2$ between $x = 0$ and $x = 5$ with 1 as the width of rectangles, using

a five left rectangles

b five right rectangles

Steps	Working
a 1 Sketch the graph. Draw the rectangles so that each touches the curve at the top left.	
2 State the left value for each rectangle.	
3 Find the height of each rectangle.	
4 Find the area of each rectangle.	
5 Add the areas.	

b 1 Sketch the graph.
Draw the rectangles so that each touches the curve at the top right.

2 State the right value for each rectangle.

3 Find the height of each rectangle.

4 Find the area of each rectangle.

SB

Using CAS 2:
Approximating
areas
p. 293

5 Add the areas.

Evaluate each definite integral.

a $\displaystyle\int_0^1 7x^3\,dx$ **b** $\displaystyle\int_3^5 \left(2x^2+1\right)dx$

Steps	Working
a 1 Integrate $7x^3$.	
2 Substitute the limits of the integral $x=1$ and $x=0$ and subtract: $F(b)-F(a)$.	
b 1 Integrate $2x^2+1$.	
2 Substitute the limits of the integral $x=5$ and $x=3$ and subtract: $F(b)-F(a)$.	

MATCHED EXAMPLE 9 | Properties of definite integrals 1

If $\displaystyle\int_1^3 g(x)\,dx = 2$, evaluate $\displaystyle\int_3^1 \big(g(x)+6\big)\,dx$.

Steps	Working
1 Simplify using Property 2: split terms.	
2 Simplify using Property 4: swap limits.	
3 Substitute $\displaystyle\int_1^3 g(x)\,dx = 2$.	
4 Evaluate and simplify.	

9780170464062

MATCHED EXAMPLE 10 | Properties of definite integrals 2

If $\int_0^1 f(x)\,dx = 4$, evaluate $\int_0^1 2\big(f(x)+3\big)\,dx$.

> Note carefully where the common factor is placed.

p. 302

Steps	Working
1 Simplify using Property 1: constant out.	
2 Simplify using Property 2: split terms.	
3 Substitute $\int_0^1 f(x)\,dx = 4$.	
4 Evaluate and simplify.	

MATCHED EXAMPLE 11 | Area under a curve

Find the area enclosed by the graph of $y = \frac{1}{2}(x-1)^2$ and the x-axis between $x = 4$ and $x = 8$.

Steps	Working
1 Sketch the graph showing the positive area required.	
2 Write the integral required to find the area.	
3 Evaluate the area.	

SB

p. 306

Find the area enclosed by the graph of $y = \frac{1}{3}x^2 - 3$ and the x-axis between $x = 1$ and $x = 4$.

Steps	Working
1 Sketch the graph, showing the area required.	
2 Find the x-intercept that splits the area into two regions, above and below the x-axis.	
3 Write the integral required to find the split area.	
4 Evaluate the integrals.	

TI-Nspire

ClassPad

MATCHED EXAMPLE 13 | Areas between curves

Find the area enclosed by the graphs of $y = x^2 + 2x - 1$ and $y = x + 1$.

Steps	Working
1 Sketch the graphs of $y = x^2 + 2x - 1$ and $y = x + 1$.	
2 Solve simultaneously to find the intersection points.	
3 Between $x = -2$ and $x = 1$, the line $y = x + 1$ is the upper function. Set up the integral using $\int_a^b (\text{upper} - \text{lower}) dx$ and simplify the terms.	
4 Integrate and evaluate.	

SB

Using CAS 5:
Area between
curves
p. 313

MATCHED EXAMPLE 14 | Average value of a function

Find the average height of an elevation rod defined by the function $g(x) = 3x^2 - 5$, between $x = 1$ and $x = 4$, where x is the distance from a certain point $(0,0)$. (Measurements are in metres.)

SB
p. 318

Steps	Working
1 State the formula for average value. average value $= \dfrac{1}{b-a}\displaystyle\int_a^b f(x)\,dx$	
2 Evaluate the integral.	
3 Use CAS to evaluate the answer.	

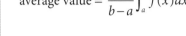

ClassPad

MATCHED EXAMPLE 15 | Differentiate, hence integrate

Find $\dfrac{d}{dx}\left(x^{-1}\right)$, and hence, find $\displaystyle\int_{2}^{3}\frac{1}{x^2}\,dx$.

Steps	Working
1 Find the derivative.	
2 Write a statement using anti-differentiation by recognition.	
3 Recognise that we need $\displaystyle\int_{2}^{3}\frac{1}{x^2}\,dx$. Multiple both sides of the equivalence by -1.	
4 Evaluate the required integral $\displaystyle\int_{2}^{3}\frac{1}{x^2}\,dx$.	
5 Use CAS to check the answer.	

> The constant of integration '$+ c$' is not needed here as we will be evaluating a definite integral.

TI-Nspire   **ClassPad**

APPLYING THE CIRCULAR FUNCTIONS

MATCHED EXAMPLE 1 | Finding the minimum and maximum value

State the minimum and the maximum value of the function $y = 3 + 4\cos[\pi(x - 1)]$ in $[0, \pi]$, and find the x values when these occur.

p. 335

Steps	Working
1 Find the range, knowing that the maximum and minimum of $\cos\theta$ are 1 and -1, respectively.	
2 Determine when the maximum and the minimum occur.	
3 Check for more solutions in the given domain.	
4 State the answer.	

MATCHED EXAMPLE 2 | Transforming the cosine function

a State the transformations required for the graph of $y = \cos(x)$ to map onto the graph of $y = -2\cos[3(x + \pi)] - 1$.

b Sketch each transformation in the interval $[0, \pi]$.

c State the range of $y = -2\cos[3(x + \pi)] - 1$ in the interval $[0, \pi]$.

Steps	Working
a 1 Compare $y = -2\cos[3(x + \pi)] - 1$ to $y = a\cos[n(x + b)] + c$, and list the values of a, b, c and n.	
2 Describe each transformation.	
b 1 Sketch $y = \cos(x)$ for $[0, \pi]$ and graph each transformation.	
2 $y = \cos(x + \pi)$	

3 $y = \cos [3(x + \pi)]$

4 $y = 2 \cos [3(x + \pi)]$

5 $y = -2 \cos [3(x + \pi)]$

6 $y = -2 \cos [3(x + \pi)] - 1$

c State the range.

9780170464062

MATCHED EXAMPLE 3 | Transforming the tangent function

The graph of $f(x) = \tan(x)$ has been transformed to the graph shown below which has equation of $g(x) = a\tan(bx) + c$, where a, b and c are constants.

a Determine the values of a, b and c.

b Describe the types of transformations applied to $f(x)$.

SB
p. 337

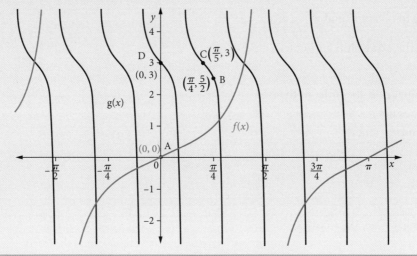

Steps	**Working**
a 1 Find the period of $g(x)$ and the value of b.	
2 Determine the value of c.	
3 Determine the value of a.	
4 State the function $g(x)$.	
b 1 Describe the effect of a on the graph of $f(x)$.	
2 Describe the effect of b on the graph of $f(x)$.	
3 Describe the effect of c on the graph of $f(x)$.	

Using CAS 1:
Graphing circular
functions
p. 338

p. 345

MATCHED EXAMPLE 4 | The product rule

Differentiate $y = \tan(x)\cos(x)$.

Steps	Working
1 Use the product rule: identify u and v.	
2 Differentiate to obtain u' and v'.	
3 Write down the expression for $uv' + vu'$ and simplify.	

> For the product rule, the order of the two functions does not matter. We can also let $u = \cos(x)$ and $v = \tan(x)$.

4 Simplify using $1 - \sin^2(x) = \cos^2(x)$	
5 State the answer.	

MATCHED EXAMPLE 5 | The quotient rule

Find $\dfrac{dy}{dx}$ for $y = \dfrac{\cos(x)}{x^3 - 1}$.

Steps	Working
1 Use the quotient rule: identify u and v.	
2 Differentiate to obtain u' and v'.	
3 Write down the expression for $\dfrac{vu' - uv'}{v^2}$ and simplify.	

> With the quotient rule the order of the two functions in the numerator *does* matter. Differentiate in alphabetical order: u first.

8

MATCHED EXAMPLE 6 | The chain rule

p. 346

Differentiate $y = \cos(\sin(x))$.

Steps	Working
1 Use the chain rule: identify u and write y in terms of u.	
2 Obtain $\dfrac{du}{dx}$ and $\dfrac{dy}{du}$ in terms of x.	
3 Use $\dfrac{dy}{dx} = \dfrac{dy}{du} \times \dfrac{du}{dx}$.	
4 State the answer.	

MATCHED EXAMPLE 7 | Combining the rules

Find $\dfrac{dy}{dx}$ for $y = \dfrac{x^4}{\cos(x^2)}$

Steps	Working
1 Identify the methods to be used.	
2 Identify u and v in the quotient rule.	
3 Obtain u' and v', that is, $\dfrac{du}{dx}$ and $\dfrac{dv}{dx}$.	
4 Use the quotient rule $dy/dx = (vu' - v'u)/v^2$.	

SB

Using CAS 2:
Differentiating
circular functions
p. 347

MATCHED EXAMPLE 8 | Equation of a tangent

Find the equation of the tangent to the curve $y = \sin(x)$ at the point where $x = \dfrac{\pi}{2}$.

Steps	Working
1 Find the y-coordinate of the point.	
2 Differentiate $y = \sin(x)$.	
3 Substitute $x = \dfrac{\pi}{2}$ into $\dfrac{dy}{dx}$ for the gradient of the tangent.	
4 Use the point–gradient formula $y - y_1 = m(x - x_1)$ with $m = 0$ and the point $\left(\dfrac{\pi}{2}, 1 \right)$.	
5 Write the answer in the general form.	

MATCHED EXAMPLE 9 | Straight-line motion

The displacement of a moving object is $x = \dfrac{1}{2}\cos(2t)$, where x is in centimetres and t is in seconds.

a Find an equation for the velocity of the object.

b What is the initial velocity of the object?

c When is the velocity first equal to zero?

d What is the maximum acceleration?

Steps	Working
a Velocity $= \dfrac{dx}{dt}$	
b Initial velocity means when $t = 0$.	
c Solve $-\sin(2t) = 0$. Only take the first solution.	
d Differentiate the velocity equation to obtain acceleration, a. Find the maximum acceleration.	

MATCHED EXAMPLE 10 | Newton's method

An object's position from a fixed point is given by the function $x = 2\cos\left(\dfrac{3\pi}{2} - t\right)$. If the object takes approximately 5.8 seconds to reach the fixed point, use Newton's method to find the first three estimates, stating the final answer correct to three decimal places.

Steps	Working
1 Find the derivative of the function.	
2 Set up the iteration process.	
3 State the first three estimates. 1st estimate 2nd estimate 3rd estimate	
4 State the answer.	

Find an anti-derivative $F(x)$ for each function.

a $f(x) = \dfrac{8}{\cos^2(2x+1)} + \sin(5x)$

b $f(x) = \sec^2(2x) - \sin(5x)$

SB

p. 357

8

Steps	Working
a 1 Anti-differentiate each term.	
2 Combine the answers and include the constant of integration.	
b 1 Anti-differentiate.	
2 State the anti-derivative and include the constant of integration.	

p. 357

MATCHED EXAMPLE 12 | Anti-differentiation by recognition

Given that $\dfrac{dy}{dx}\left((x^2+1)\sin x\right)=2x\sin(x)+(x^2+1)\cos(x)$, write the expression for

$\displaystyle\int\left[3(x^2+1)\cos(x)+6x\sin(x)\right]dx$.

Steps	Working
1 Write the derivative as an integral.	
2 Write the required answer in terms of the known integral.	
3 State the answer. Remember to add the constant of integration.	

MATCHED EXAMPLE 13 | Anti-differentiating to find the constant of integration

The derivative of a function is $f'(x) = \dfrac{2}{\cos^2(2x)}$. If $f\left(\dfrac{\pi}{8}\right) = -\dfrac{1}{8}$, determine the function $f(x)$.

Steps	Working
1 Anti-differentiate $f'(x) = \dfrac{2}{\cos^2(2x)}$.	
2 Find the value of c using $f\left(\dfrac{\pi}{8}\right) = -\dfrac{1}{8}$.	
3 State the answer.	

MATCHED EXAMPLE 14 | Evaluating definite integrals

Evaluate each definite integral.

a $\int_0^{\frac{\pi}{6}} \sec^2(2x)\,dx$

b $\int_{-\pi}^{\frac{\pi}{2}} \cos\left(\frac{4x}{3}\right)\,dx$

Steps	Working
a Integrate the function and substitute the limits of integration.	
b Integrate the function and substitute the limits of integration.	

Find the exact area under the graph of $y = \dfrac{1}{\sqrt{2}}\cos\left(\dfrac{x}{4}\right)$ between $x = 0$ and $x = \pi$, shown by the shaded region.

SB

p. 362

Steps	Working
1 Write the area as a definite integral.	
2 Evaluate the integral.	
3 Write the answer.	

p. 363

MATCHED EXAMPLE 16 | Areas above and below the *x*-axis

a Sketch the graph of the curve $y = 3\sin(2x)$ in the interval $[-\pi, \pi]$.

b Calculate the area bounded by the curve and the *x*-axis in the interval $\left[-\pi, \dfrac{\pi}{2}\right]$.

Steps	Working
a Sketch the graph in the interval $[-\pi, \pi]$.	
b **1** Write the definite integrals for the area, using a negative sign for the region below the *x*-axis.	
2 Calculate the total area.	
We can save time if we recognise that the areas of each section are the same.	
3 Answer the question.	

Using CAS 5:
Area between two
curves
p. 364

CHAPTER 9

APPLYING THE EXPONENTIAL AND LOGARITHMIC FUNCTIONS

SB

p. 386

MATCHED EXAMPLE 1	Using transformations to sketch an exponential function
a Sketch the graph of $f(x) = e^{2x}$	
b Using mapping of points, sketch the graph of $2 - 5f(3x + 1)$	

Steps	Working
a Sketch the graph of $f(x) = e^{2x}$	
b **1** Identify three coordinates to use when mapping.	
2 Compare the function with the general exponential function $y = e^x$ and identify each transformation. Use this information to sketch the graph.	
3 Write out the mapping sequence in order of transformation.	

4 Use mapping of points to find new points.

5 Consider the new asymptote $y = 2$. Use the new points and the asymptote $y = 2$ to sketch the graph.

Using CAS 1: Graphing exponential functions p. 387

9780170464062

SB

p. 387

Sketch the graph of $y = 5e^x - 2$, showing the coordinates of any axial intercepts and labelling the equation of the asymptote.

Steps	**Working**
1 Compare the function with the general exponential function $y = e^x$ and identify each transformation.	
2 Consider the new y-intercept and the new x-intercept. Use this information to sketch the graph.	

9

SB

p. 388

MATCHED EXAMPLE 3 | Finding the rule of an exponential function

The rule for the function with the graph shown is of the form $y = ae^x + b$.

a Find the values of a and b.

b Hence, state the equation of the function.

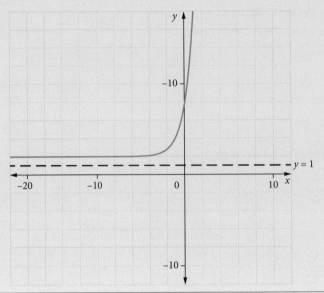

Steps	Working
a 1 Using the horizontal asymptote, find the value of b.	
2 Substitute $(0, 7)$ into the general equation.	
b Write the equation of the function.	

9780170464062

MATCHED EXAMPLE 4	Using simultaneous equations to find the rule of an exponential function	

Given that $y = Ae^{bt}$ with $y = 5$ when $t = 2$ and $y = 15$ when $t = 3$, find the value of A and b.

p. 388

Steps	Working
1 Form two equations in A and b by substituting into the rule $y = Ae^{bt}$.	
2 Divide equation [2] by equation [1].	
3 Substitute $\log_e (3) = b$ into equation [1].	
4 Write the rule.	

MATCHED EXAMPLE 5 | Modelling an exponential decay problem

The difference between the temperature of a cup of hot chocolate and its surroundings decreases by 3% every minute. A fresh cup has a temperature of 80°C, and the room temperature is 18°C. A cup of hot chocolate is considered too cold to drink if it is below 30°C. Model the temperature of the cup of hot chocolate and find how long (to the nearest minute) it takes to become undrinkable.

Steps	Working
1 This is an example of exponential decay.	
2 Write the relationship between the functions. Room temperature = 18°C.	
3 Write $D(m)$ as an exponential function. $D(m)$ decreases by 3% each minute, so $a = 97\% = 0.97$. The initial difference was $80° - 18°$.	
4 Make $C(m)$ the subject.	
5 To find the time taken for the hot chocolate to become undrinkable, we want $C(m) = 30$. Try some different values of m until $C(m)$ is close to 30.	
6 Answer the question.	

MATCHED EXAMPLE 6 | Modelling an exponential growth problem

SB

p. 392

The number N of trees with cherries in a region of Northern Australia was studied.

The equation $N = 500e^{0.05t}$ was given as a model for the increase, with time t measured in years.

a How many trees were there at the beginning of the study?

b How many trees were there after 7 years?

Steps	Working
a 1 This is an example of exponential growth. Substitute $t = 0$ into $N = 500e^{0.05t}$.	
2 Answer the question.	
b 1 Substitute $t = 7$ into $N = 500e^{0.05t}$.	
2 Answer the question.	

MATCHED EXAMPLE 7 Finding the derivative of an exponential function

Differentiate each exponential function.

a $y = e^{7x+1}$

b $y = e^{2x^2 - x + 3}$

Steps	Working
a Use the rule $\dfrac{d}{dx}(e^{ax}) = ae^{ax}$.	
b 1 Use the chain rule: identify u and write y in terms of u.	
2 Write $\dfrac{du}{dx}$ and $\dfrac{dy}{dx}$ in terms of x.	
3 Use $\dfrac{dy}{dx} = \dfrac{dy}{du} \times \dfrac{du}{dx}$.	

MATCHED EXAMPLE 8 | The chain and product rules

p. 396

Find the derivative of each of the following functions.

a $y = \cos(2e^{3x})$

b $y = x^3 e^{-5x}$

Steps	Working
a 1 Use the chain rule: identify u and write y in terms of u.	
2 Obtain $\dfrac{du}{dx}$ and $\dfrac{dy}{du}$ in terms of x.	
3 Use $\dfrac{dy}{dx} = \dfrac{dy}{du} \times \dfrac{du}{dx}$.	
b 1 Use the product rule: identify u and v.	
2 Differentiate to obtain u' and v'.	
3 Write down the expression for $uv' + vu'$ and simplify.	

p. 397

MATCHED EXAMPLE 9 The quotient rule

Find the derivative of $f(x) = \sqrt{\dfrac{e^{5x}}{2x^3 + 2x + 5}}$ at the value where $x = 1$.

Steps	Working
1 Use the quotient rule: identify u and v.	
2 Differentiate to obtain u' and v'.	
3 Write down the expression for $\dfrac{vu' - uv'}{v^2}$ and simplify.	
4 Use the chain rule: identify u and write y in terms of u.	
5 Use chain rule to obtain $f'(x)$.	
6 Find $f'(1)$ by substituting $x = 1$ into $f'(x)$.	

Using CAS 3:
Finding the
derivatives of
functions involving
exponentials
p. 397

Find the equation of the tangent to the curve $f(x) = \dfrac{e^{\sqrt{3x+1}}}{\cos(2x)}$ at $x = 2$.

SB
p. 398

Steps	Working
1 Use the quotient rule: identify u and v.	
2 Differentiate to obtain u' and v'.	
3 Write down the expression for $\dfrac{vu' - uv'}{v^2}$ and simplify.	
4 Find the gradient of the tangent at $x = 2$.	
5 $f(2)$	
6 Find the 'c' of the tangent line by substituting m_T and the point into the linear equation $y = mx + c$.	
7 Use $y = mx + c$ to find the equation of the tangent.	

SB

Using CAS 4:
Finding the
equation of the
tangent
p. 399

MATCHED EXAMPLE 11 | Using the natural exponential function and its derivative

A piece of brass cools down according to the formula $T = T_0 e^{-0.5t}$, where T is the temperature difference between the metal and the surroundings in °C and t is in minutes. The initial temperature is 918°C and the room is at 18°C.

a Evaluate T_0.

b Find correct to one decimal place the temperature difference after 10 minutes.

c What is the temperature after 10 minutes? Answer correct to one decimal place.

d Find correct to one decimal place the rate at which the piece of brass is cooling after 10 minutes.

Steps	Working
a Find the initial temperature difference with the surroundings.	
b 1 Substitute $t = 10$ into $T = T_0 e^{-0.5t}$.	
2 State the result.	
c Add the room temperature.	
d 1 Find the derivative.	
2 Substitute $t = 10$.	
3 State the result.	

9780170464062

MATCHED EXAMPLE 12 | Using the derivative with exponential decay

Radon-226 decays at a continuous rate of 15.2% per day.

a How much, correct to three decimal places, will 150 mg of Radon-226 decay to in 5 days?

b At what rate is Radon-226 decaying after 2 weeks?

Steps	Working
a 1 Use the continuous growth formula.	
2 Substitute $a = 150$, $r = -0.152$ and $t = 5$ into the continuous growth formula. The substance is decaying, so we know the growth rate (r) will be negative.	
3 State the result.	
b 1 Substitute $a = 150$ and $r = -0.152$ into the continuous growth formula.	
2 In order to find the rate, find $f'(t)$.	
3 Substitute $t = 14$ into the derivative of the continuous growth formula.	
4 State the result.	

SB

p. 405

MATCHED EXAMPLE 13 | Finding the integral of an exponential function

Find the integral of each of the following:

a $\displaystyle\int e^{5-x}\,dx$ **b** $\displaystyle\int \frac{e^{2x}+7+e^{6x}}{3e^{5x}}\,dx$

Steps	Working
a Use $\displaystyle\int e^{ax+b}\,dx = \frac{1}{a}e^{ax+b}+c$ with $a=-1$, $b=5$.	
b **1** Separate the terms first.	
2 Simplify by subtracting powers.	
3 Integrate each term using $\displaystyle\int e^{ax}\,dx = \frac{1}{a}e^{ax}+c$	

MATCHED EXAMPLE 14 | Finding $f(x)$ given $f'(x)$ and a point

Find $f(x)$ if its gradient function is $f'(x) = 2e^{5x}$ and $f(0) = 3$.

Steps	Working
1 Integrate $f'(x)$ to find $f(x)$.	
2 Substitute $x = 0$, $f(x) = 3$ to find c.	
3 Write the equation of the function.	

9

MATCHED EXAMPLE 15 | Evaluating definite integrals

Evaluate $\int_{-1}^{1} 2e^{-\frac{2x}{3}} + 3 \cos(5x)\ dx$ correct to two decimal places.

Steps	Working
1 Integrate each term separately.	
2 Evaluate the integral.	

MATCHED EXAMPLE 16 | Using transformations to sketch a logarithmic function

a Sketch the graph of $f(x) = \log_e (x)$.

b Using mapping of points, sketch the graph of $3 - 2f(5 - x)$.

Steps	Working
a Sketch the graph of $f(x) = \log_e (x)$.	
b 1 Identify three coordinates to use when mapping.	
2 Compare the function with the general exponential function $f(x) = \log_e (x)$ and identify each transformation.	
3 Write out the mapping sequence in order of transformation occurrence.	
4 Use mapping of points to find new points.	

5 Consider the new asymptote $x = 5$. Use the new points and the asymptote $x = 5$ to sketch the graph.

MATCHED EXAMPLE 17	Using the intercept method to sketch a logarithmic function

Sketch the graph of $f(x) = \log_e(x) + 5$, labelling important features.

SB

p. 411

Steps	Working
1 State the translation.	
2 Find the x-intercept by solving $f(x) = 0$.	
3 State the asymptote.	
4 Since there is no y-intercept, state another point.	
5 Sketch the graph.	

SB

Using CAS 6:
Graphing
logarithmic
functions
p. 411

MATCHED EXAMPLE 18 Finding the rule of a logarithmic function

The rule for the function with the graph shown is of the form $y = a \log_e (x - b)$. Find the values of a and b. Hence, state the equation of the function.

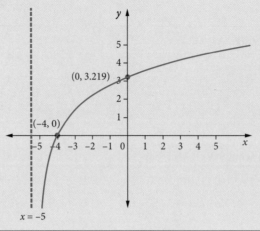

Steps	Working
1 Substitute $(0, 3.219)$ into the general equation.	
2 Substitute $(-4, 0)$ into the general equation.	
3 Name the equations as equation [1] and [2] to be solved simultaneously.	
4 To eliminate a, perform $[2] \div [1]$.	
5 Solve for b.	
6 Substitute $b = -5$ into [1] to find a.	
7 Write the equation.	

MATCHED EXAMPLE 19 The chain rule

Differentiate each logarithmic function using the chain rule.

a $y = \log_e (7x + 1)$ **b** $y = \log_e (3x - 2)^2$

c $y = \sqrt{\log_e (2x^3)}$ **d** $y = \log_5 (x - 1)$

Steps	Working
a 1 Use the chain rule: identify u and write y in terms of u.	
2 Write $\dfrac{du}{dx}$ and $\dfrac{dy}{du}$ in terms of x.	
3 Use the chain rule $\dfrac{dy}{dx} = \dfrac{dy}{du} \times \dfrac{du}{dx}$.	
4 State the answer.	
b 1 Given $g(x) = (3x - 2)^2$, find $g'(x)$.	
2 Use $\dfrac{dy}{dx} = \dfrac{g'(x)}{g(x)},\ g(x) \neq 0$ to differentiate and simplify.	
Or using logarithm laws:	
1 Write $y = \log_e (3x - 2)^2$ using the logarithmic law $\log_a(x^n) = n\log_a(x)$.	
2 Use $\dfrac{d}{dx}\big(\log_e (ax + b)\big) = \dfrac{a}{ax + b}$ to differentiate and simplify.	
c 1 Rewrite y and change from radical form to power form.	
2 Use the chain rule: identify u and write y in terms of u.	
3 Write $\dfrac{du}{dx}$ and $\dfrac{dy}{du}$ in terms of x. Notice that $\dfrac{du}{dx}$ is found by using the chain rule $\dfrac{du}{dx} = \dfrac{du}{dw} \times \dfrac{dw}{dx}$ and $\dfrac{dy}{dx}$ is also found by using the chain rule. So, we are using a chain rule within a chain rule.	

4 Use $\dfrac{dy}{dx} = \dfrac{dy}{du} \times \dfrac{du}{dx}$.

d 1 Use the change of base formula

$\log_b(x) = \dfrac{\log_a(x)}{\log_a(b)}$ with $b = 5$ to write y in

terms of $\log_e(x)$.

> Note: $\dfrac{1}{\log_e(5)}$ is a constant (number).

SB

Using CAS 7:
Differentiating
logarithmic
functions
p. 417

2 Use $\dfrac{d}{dx}(\log_e x) = \dfrac{1}{x}, x \neq 0$ to differentiate and
simplify.

MATCHED EXAMPLE 20	The product rule

Find the derivative of $f(x) = 2x^3 \ln (5x)$.

Steps	Working
1 Use the product rule: identify u and v.	
2 Differentiate to obtain $\dfrac{du}{dx}$ and $\dfrac{dv}{dx}$.	
3 Write down the expression for $\dfrac{dy}{dx} = v\dfrac{du}{dx} + u\dfrac{dv}{dx}$ and simplify.	

9

MATCHED EXAMPLE 21 | Finding the equation of the tangent

Find the equation of the tangent to the curve $f(x) = \log_e(3x^2)$ at

a $x = 1$

b $x = a$

Steps	Working
a 1 Differentiate $f(x)$.	
2 Find $f'(1)$ and $f(1)$.	
3 Use the point–gradient formula to find the equation of the tangent.	
b Find $f'(a)$ and $f(a)$.	

Find the equation of the tangent to the curve $f(x) = \log_e(3x^2)$ at

a $x = 1$

b $x = a$

MATCHED EXAMPLE 22 | Integrating $\dfrac{1}{ax+b}$

Find $\displaystyle\int \dfrac{3}{5-2x}\,dx$, with an appropriate restriction of the domain.

Steps	Working
1 Factorise by the constant 3.	
2 Use $\displaystyle\int \dfrac{1}{ax+b}\,dx = \dfrac{1}{a}\log_e(ax+b)+c$ with $a=-2$.	
3 Restrict the domain to $5-2x>0$ for log.	
4 Write the answer.	

MATCHED EXAMPLE 23 | Evaluating definite integrals

Evaluate

a $\displaystyle\int_{1}^{2}\frac{dx}{5x-3}$

b $\displaystyle\int_{1}^{3}\frac{6x-2}{3x+1}dx$

Steps	Working
a 1 Find the integral.	
2 Evaluate the integral.	
b 1 Simplify $\dfrac{6x-2}{3x+1}$ into $\dfrac{a}{x+b}+c$.	
2 Find the integral.	

Find the equation of the curve $f(x)$ given that $f'(x) = \dfrac{3}{7-2x}$ and the curve passes through $(2, 0)$.

SB

p. 421

Steps	Working
1 Integrate $f'(x)$ to find $f(x)$.	
2 Substitute $(2, 0)$ into $f(x)$ to find c.	
3 Write the equation.	

9

p. 422

MATCHED EXAMPLE 25 | Integrating by recognition 1

Differentiate $f(x) = x^2 \log_e x$. Hence, find the indefinite integral $\int 2x \log_e (x)\, dx$.

Steps	Working
1 To differentiate $x^2 \log_e x$, use the product rule: identify u and v.	
2 Find u' and v' in terms of x.	
3 Use the product rule. $vu' + uv'$	
4 Differentiate $-\dfrac{1}{2}x^2$.	
5 Add this to $f'(x)$ to give the expression to be integrated.	
6 Write a concluding statement and simplify.	

Given that $\int_{1}^{k}\left(\dfrac{5}{5x+1}\right)dx=9$, find the value of k.

SB

p. 422

Steps	Working
1 Find the integral.	
2 Make this equal to 9 and solve.	

MATCHED EXAMPLE 27	Integral of $\dfrac{f'(x)}{f(x)}$

Find each integral.

a $\displaystyle\int \frac{3x^2+1}{x^3+x-2}\,dx$ **b** $\displaystyle\int \frac{16x}{4x^2-3}\,dx$

Steps	Working
a 1 Test if the numerator is the derivative of the denominator.	
2 It is, so the integral is a logarithmic function. $\displaystyle\int \frac{f'(x)}{f(x)}\,dx = \log_e\left[f(x)\right]+c$	
b 1 Test if the numerator is the derivative of the denominator.	
2 It almost is, so adjust the integral.	

MATCHED EXAMPLE 28 | Integration by recognition 2

Find $\dfrac{d}{dx}$ [ln $(2x^2 + 7)$]. Hence, find $\displaystyle\int_1^4 \dfrac{x}{2x^2 + 7}\, dx$.

Steps	Working
1 To differentiate $\log_e (2x^2 + 7)$, identify u and y.	
2 Find u' and y'.	
3 Use chain rule, $\dfrac{dy}{du} \times \dfrac{du}{dx}$.	
4 Write a 'hence case' statement.	
5 Alter the hence case statement to suit the problem.	
6 Rearrange the hence case statement and evaluate the answer.	

MATCHED EXAMPLE 29 | Newton's method

Use Newton's method to approximate a zero for the cubic function $y = \log_e (5x - 3)$.

If $x_0 = 1$, find the value of x_1.

Steps	Working
1 Differentiate $y = \log_e (5x - 3)$.	
2 Calculate $f(x_0)$ and $f'(x_0)$ where $x_0 = 1$.	
3 Use Newton method to find x_1. $x_{n+1} = x_n - \dfrac{f(x_n)}{f'(x_n)}$	

What is the *exact* value of this zero? You can solve this algebraically.

| MATCHED EXAMPLE 30 | Modelling with logarithmic functions |

SB

p. 428

Amy is learning to speak French before going to New Caledonia for 10 months. While completing an online course, the number of words he learns, W, is modelled by the function

$$W = 90 \log_e (t - 1) + 110$$

where t is the time in days.

Amy needs a very basic vocabulary of 500 words for the trip, which will be in four months' time.

a How long will it take him to learn the very basic vocabulary?

b Find an expression for his rate of learning.

c How long would it take before his rate of learning dropped to one word per day?

Steps	Working
a 1 Substitute $W = 500$ and solve to find t.	
2 Answer the question.	
b Write the expression for the rate of learning.	
c 1 Find when $\dfrac{dW}{dt} = 1$.	
2 Answer the question.	

DISCRETE PROBABILITY AND THE BINOMIAL DISTRIBUTION

SB

p. 445

MATCHED EXAMPLE 1	Finding probabilities using a Venn diagram

If $\Pr(A)=\dfrac{2}{3}$, $\Pr(B)=\dfrac{3}{7}$ and $\Pr(A\cap B)=\dfrac{1}{7}$, use a Venn diagram to find $\Pr(A'\cap B')$.

Steps	Working
1 Express all fractions with a common denominator.	
2 Illustrate the probabilities with a Venn diagram. Fill the centre (intersection) first.	
3 Complete the remaining part of set A and set B.	Answer in the diagram you drew in step 1.
4 $(A'\cap B')$ is the part of the sample space that is NOT in set A or set B.	Answer in the diagram you drew in step 1.

9780170464062

For two events A and B, $\Pr(A \cap B) = p$, $\Pr(B) = 0.56$, $\Pr(A) = p + 0.12$ and $\Pr(A' \cap B') = 2p$. Find $\Pr(A \cap B')$.

SB

p. 446

Steps	Working
1 Enter the given probabilities into a probability table.	
2 $\Pr(A \cap B')$ and $\Pr(B')$ can be found first.	Write your answer in the table you drew in step 1.
3 Find the value of p from the B' column. Then, substitute into the table and calculate the remaining probabilities.	
4 Find $\Pr(A \cap B')$ in the table.	

MATCHED EXAMPLE 3 | Finding probabilities for independent events

For two events A and B, $\Pr(A \cap B) = 0.15$, $\Pr(B) = p$ and $\Pr(A') = 0.8 - p$. Find the value of p if A and B are independent.

Steps	Working
1 Events A and B are independent. Write the independence rule.	
2 Find $\Pr(A)$ using the complementary probability $\Pr(A')$.	
3 Substitute into the independence rule. Solve for p.	

9780170464062

A box contains 18 pencils, of which 6 are blue coloured. Two pencils are randomly selected from the box, **without replacement**. This means that a pencil is selected and the pencil is **not** replaced **before** the next pencil is selected. Find the probability that one of the two pencils is blue coloured.

SB
p. 447

10

Steps	Working
1 On each selection, the pencil selected can be either blue coloured (B) or not blue coloured (B').	
The first selection is made from 6 pencils that are blue coloured and 12 pencils that are not blue coloured.	
In the second selection, the number of blue-coloured and not blue-coloured pencils is determined by the type of pencil selected first.	
2 Represent with a tree diagram.	
3 Identify the branches where there is one blue-coloured pencil and one pencil that is not blue coloured.	
4 Multiply the probabilities along the branches and add the products.	

SB

Using CAS 1:
Selection without
replacement
probabilities
p. 448

MATCHED EXAMPLE 5 Finding probabilities for dependent events

On Mondays, Claire has coffee at either Monday's Coffee Store or The Cupping Room. If Claire goes to Monday's Coffee Store one week, there is a probability of 0.6 she will go to The Cupping Room the following week. If she goes to The Cupping Room, there is a probability of 0.9 she will go to Monday's Coffee Store the following week. This Monday, Claire went to Monday's Coffee Store. Find the probability that she will go to The Cupping Room the next two Mondays.

Steps	Working
1 Set up the tree diagram for the next two Mondays. M represents coffee at Monday's Coffee Store. C represents coffee at The Cupping Room.	
2 Identify the branches where Claire has coffee at The Cupping Room the next two Mondays, and multiply the probabilities.	

p. 449

If $\Pr(A \cup B) = \dfrac{3}{7}$, $\Pr(A \cap B) = \dfrac{1}{7}$ and $\Pr(B) = \dfrac{3}{2} \times \Pr(A)$, find $\Pr(B)$.

Steps	Working
1 Let $\Pr(A) = a$.	
Write $\Pr(B)$ in terms of a.	
2 Substitute in the addition rule.	
3 Substitute the value of a into $\Pr(B) = \dfrac{3}{2}\, a$.	

MATCHED EXAMPLE 7 | Finding conditional probabilities

Mechanic Ben owns a car, a jeep and an electric bicycle. Each morning, he starts each vehicle, but they do not always start on the first attempt. The respective probabilities of each starting on the first attempt are 0.3, 0.6 and 0.8. On Monday morning, two vehicles started on the first attempt. Find the probability the two that started were the jeep and the electric bicycle.

Steps	Working
1 Write the conditional probability in the form $$\Pr(A|B) = \frac{\Pr(A \cap B)}{\Pr(B)}.$$	
2 Find the probability that the car and the jeep start but the electric vehicle does not start. Let C = car starts, J = jeep starts and E = electric bicycle starts.	
3 Find the probability that two vehicles start, that is, $\Pr(CJE') + \Pr(CJ'E) + \Pr(C'JE)$.	
4 Substitute into the conditional probability formula.	

MATCHED EXAMPLE 8 | Finding the probability distribution of a discrete random variable

SB

p. 456

A pencil case contains fifteen pens of which nine are blue. Two pens are taken from the pencil case. If X represents the number of blue pens taken, list the probability distribution of X.

Steps	Working
1 Draw a tree diagram to illustrate the problem.	
2 Draw a probability distribution table. X represents the number of blue pens selected and $x \in \{0, 1, 2\}$.	
3 Use the tree diagram to find the probabilities.	
4 Write the probabilities in the table.	

p. 456

MATCHED EXAMPLE 9 Finding a probability for a given probability distribution

The probability distribution of a discrete random variable, X, is given by the table below.

Find the value of p.

x	1	2	3	4
$\Pr(X = x)$	$2p^2 - 0.6$	$p - 0.2$	$p^2 - 0.4$	0.2

Steps	**Working**
1 As this is a probability distribution, the sum of the probabilities is one.	
2 Factorise and solve for p.	
3 As p is a probability, $p \geq 0$.	

SB
p. 457

The Australian newspaper publishes three sudokus every day. Jen attempts them each day and has calculated the probability of solving 0, 1, 2 or 3 sudokus. The probability distribution of the number of sudokus solved, X, is given in the table below.

Find the probability that Jen solves the same number of sudokus on Friday and Saturday.

x	0	1	2	3
$\Pr(X = x)$	0.1	0.6	0.1	0.2

Steps	Working
1 List the possible options where Jen solves the same number of sudokus on two days. $\Pr(0, 0)$ is the probability of 0 puzzles solved on Friday and on Saturday.	
2 Find the probabilities from the table, and multiply the probabilities for each pair.	

MATCHED EXAMPLE 11 | Finding the expected value of a discrete probability distribution

The probability distribution of a discrete random variable, X, is shown below.

x	0	1	2	3
$\Pr(X = x)$	0.1	0.3	0.1	0.5

a Find the mean of X.

b Find $\Pr(X < \mu)$.

Steps	Working
a 1 Rewrite the table with rows as columns and add a column headed $x \times p(x)$. Calculate the product of the x values and their probabilities. **2** The sum of the $x \times p(x)$ column is the expected value of X or $E(X)$.	
b 1 Identify the x values in the table that are less than 2 and add their probabilities.	

MATCHED EXAMPLE 12	Finding probabilities given the expected value of a discrete probability distribution

SB

p. 460

The probability distribution of a discrete random variable, X, is shown below.

x	1	2	3	4
$\Pr(X = x)$	0.1	a	0.4	b

The mean of the distribution is 3.1. Find the values of a and b.

Steps	**Working**
1 Rewrite the table with rows as columns and add a column headed $x \times p(x)$, and calculate the product of the x values and their probabilities.	
2 The sum of the $p(x)$ column must be 1. The sum of the $x \times p(x)$ column must be $E(X) = 3.1$.	
3 Write the equations for the total of the $p(x)$ and $x \times p(x)$ columns.	
4 Solve the simultaneous equations [1] and [2] to find the values of a and b.	

MATCHED EXAMPLE 13 Finding the mode of a discrete probability distribution

The probability distribution of a discrete random variable, X, is shown below.

Find the mode.

x	5	6	7	8
$\Pr(X = x)$	0.4	0.18	0.17	0.25

Steps	Working
$X = 5$ has the highest probability, 0.4.	

MATCHED EXAMPLE 14	Finding the variance and standard deviation of a discrete probability distribution

For the probability distribution below, find

x	0	1	2	3	4
$p(x)$	0.1	0.2	0.3	0.3	0.1

a the expected value

b the variance

c the standard deviation, correct to three decimal places.

Steps	Working
a 1 Add two extra columns headed $x \times p(x)$ and $x^2 \times p(x)$ for calculating $E(X)$ and $E(X^2)$, respectively. **2** Multiply x by $p(x)$, enter the results in the $x \times p(x)$ column and find the total. **3** Multiply x by $x \times p(x)$, enter the results in the $x^2 \times p(x)$ column and find the total. **4** The total of the $x \times p(x)$ column is $E(X)$.	
b 1 The total of the $x^2 \times p(x)$ column is $E(X^2)$. **2** Use the computational formula to find $\text{Var}(X)$.	
c 1 Find the standard deviation using this formula: $SD(X) = \sqrt{\text{Var}(X)}$.	

SB

p. 463

SB

Using CAS 2:
Expected value,
variance and
standard deviation
p. 463

SB

Using CAS 3:
The binomial
distribution
p. 469

MATCHED EXAMPLE 15 Finding the expected value and variance of $aX + b$

A discrete random variable, X, has an expected value of 15.7 and a variance of 3.4.

Find

a $E(5X + 1)$ **b** $Var(5X + 1)$

Steps	Working
a Use the formula $E(aX + b) = a\,E(X) + b.$	
b Use the formula $Var(aX + b) = a^2\,Var(X).$	

MATCHED EXAMPLE 16 | Finding binomial probabilities

In a basketball match, a shooting guard has a probability of 0.4 of scoring a goal from a free throw line. If he is given a chance for 7 free throws for each goal in a match, find the probability, correct to three decimal places, that he shoots

a 2 goals. **b** at least 2 goals.

Steps	Working
a 1 The distribution is binomial because there are two outcomes on each trial: shot a goal or did not shoot a goal.	
2 Write the probability using this formula: $$\Pr(X=x) = {}^nC_x \, p^x \, (1-p)^{n-x}$$	
3 Use CAS to calculate the answer. Use the binomial probability density function BinomialPdf (TI-Nspire) or BinomialPDf (ClassPad) as the probability is a single outcome.	
b 1 Write the required probability.	
2 Use CAS to calculate the answer. Use the binomial cumulative distribution function BinomialCdf (TI-Nspire) or BinomialCDf (ClassPad) as the probability is a range of outcomes. The lower bound is 2 and the upper bound is 7. These bounds are inclusive.	

TI-Nspire

ClassPad

Select **Interactive** > **Distribution/Inv.Dist** > **Discrete** > **binomialCDf**.

Select **menu** > **Probability** > **Distributions** > **Binominal Cdf**.

3 Write the probability correct to three decimal places.

p. 471

MATCHED EXAMPLE 17 | Using the binomial distribution formula

A biased coin is tossed seven times. The probability of a head occurring on any toss is p.

a Find, in terms of p, the probability of obtaining

  i seven heads.   **ii** six heads.

b Find the value of p if the probability of seven heads is equal to the probability of six heads.

Steps	Working

a   **i**   **1**   The distribution is binomial because there are two possible outcomes on each trial: a head or a tail.
 	
     2   Calculate the probability of $X = 7$ using the formula
       $\Pr(X = x) = {}^{n}C_{x}\, p^{x}(1-p)^{n-x}$.
 	
  ii   Calculate the probability of $X = 6$ using the formula
       $\Pr(X = x) = {}^{n}C_{x}\, p^{x}(1-p)^{n-x}$.	

b Equate the answers in part **a** and solve for p.	

MATCHED EXAMPLE 18	Finding the mean and variance of a binomial distribution

SB

p. 472

A binomial random variable has a mean of 20 and a variance of 12. Find the values of n and p.

Steps	Working
1 $E(X) = np$	
$\text{Var}(X) = np(1 - p)$	
2 Solve the equations by substitution to find the values of n and p.	

SB

Using CAS 4:
Finding the value
of n for a binomial
distribution
p. 472

MATCHED EXAMPLE 1	Sketching a continuous probability density function

The probability density function for a continuous random variable X is given by the piecewise function

$$f(x)=\begin{cases} -\dfrac{x}{9} & -3\le x<0 \\[2mm] \dfrac{x}{9} & 0\le x\le 3 \\[2mm] 0 & \text{elsewhere} \end{cases}$$

Sketch this probability density function.

Steps	**Working**

1 Determine the three rules that define the hybrid
  function and their domains.

2 Sketch the graphs over the four domains.

MATCHED EXAMPLE 2 | Finding a parameter in a continuous probability density function

The probability density function for a continuous random variable X is given by

$$f(x) = \begin{cases} ax+1 & 0.1 \le x \le 0.5 \\ 0 & \text{otherwise} \end{cases}$$

Find the value of a.

Steps	Working
1 Find the value of the endpoints of the line and then sketch the pdf.	
2 Use the formula for the area of a trapezium: $A = \dfrac{1}{2}(a+b)h$ 	
3 The area under the probability density function is equal to 1.	

p. 490

MATCHED EXAMPLE 3 | Integrating to find a parameter in a probability density function

The probability density function for a continuous random variable X is given by

$$f(x) = \begin{cases} ax(2x^2 - 3) & 0 \le x \le 2 \\ 0 & \text{otherwise} \end{cases}$$

Find the value of a.

Steps	Working
1 Find an expression for the total area under the probability density function by integration.	
2 This area equals 1.	

Using CAS 1:
Solving integral
equations
p. 491

The probability density function for a continuous random variable X is given by

$$f(x) = \begin{cases} ax(2x^2 - 3) & 0 \le x \le 2 \\ 0 & \text{otherwise} \end{cases}$$

MATCHED EXAMPLE 4   Finding probabilities in a continuous probability density function 1

The probability density function for a continuous random variable X is given by

$$f(x) = \begin{cases} \dfrac{x}{24} & 1 \leq x \leq 7 \\ 0 & \text{otherwise} \end{cases}$$

Find $\Pr(X < 4)$.

Steps	**Working**

1 Find the values of the endpoints of the line and then sketch the pdf.

> Remember to include the zeros and identify the nature of the endpoints of the graph.

2 Shade the area for $\Pr(X < 4)$.

To find this area, use the formula for the area of a trapezium:

$$A = \frac{1}{2}(a+b)h$$

Alternatively, this area can be calculated by integration.

MATCHED EXAMPLE 5 Finding probabilities in a continuous probability density function 2

The probability density function for a continuous random variable X is given by

$$f(x) = \begin{cases} x & 0 \le x < 1 \\ 2-x & 1 \le x \le 2 \\ 0 & \text{otherwise} \end{cases}$$

Find $\Pr(X < \frac{3}{2})$.

Steps	Working
1 The probability density function has two linear functions. Find the value of the endpoints of the lines and then sketch the pdf. Shade in the area indicated by $\Pr\left(X < \dfrac{3}{2}\right)$.	
2 Express the probability as a definite integral.	
3 The first integral is the area of a triangle with base 1 and height 1, so it is equal to $\dfrac{1}{2}$.	

MATCHED EXAMPLE 6 Finding conditional probabilities for continuous distributions

The lifetime of an appliance, in days, is a continuous random variable that is modelled by the probability density function

$$f(t) = \begin{cases} 0.5e^{-0.5t} & t \geq 0 \\ 0 & \text{otherwise} \end{cases}$$

If a randomly selected appliance lasts at least 80 days, find the probability that the appliance lasts fewer than 120 hours.

Steps	Working
1 Write a mathematical expression for the conditional probability.	
2 Find $\Pr(T < 120) \cap \Pr(T \geq 80)$. 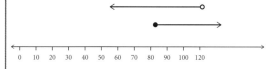	
3 Find $\Pr(T \geq 80)$.	
4 Find the conditional probability.	

MATCHED EXAMPLE 7	Finding the expected value of a continuous probability density function

The probability density function for a continuous random variable X is given by

$$f(x) = \begin{cases} \dfrac{5}{32}x^4 & \text{if } 0 \le x \le 2 \\ 0 & \text{elsewhere} \end{cases}$$

Find the expected value of X.

Steps	Working
1 Write the integral for the formula $E(X) = \int_{-\infty}^{\infty} x f(x)dx$ and simplify.	
2 Evaluate the integral.	

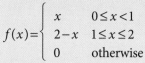

MATCHED EXAMPLE 8 | Finding the expected value of a piecewise continuous PDF

p. 501

The probability density function for a continuous random variable X is given by

$$f(x)=\begin{cases} x & 0\le x<1 \\ 2-x & 1\le x\le 2 \\ 0 & \text{otherwise} \end{cases}$$

Find $E(X)$.

Steps	Working
1 Write the two integrals for the formula $E(X)=\int_{-\infty}^{\infty}xf(x)dx$ and simplify.	
2 Evaluate the integrals.	

Using CAS 3:
Expected value
p. 502

MATCHED EXAMPLE 9 | Finding a percentile of a continuous probability density function

The probability density function for a continuous random variable X is given by

$$f(x)=\begin{cases} \dfrac{x}{18} & 0 \le x \le 6 \\ 0 & \text{elsewhere} \end{cases}$$

Find the value of a such that $\Pr(X < a) = 0.25$.

Steps	Working
1 Graph the probability density function.	
2 Find the value of a by solving $\int_{-\infty}^{a} f(x)\,dx = 0.25$	

9780170464062

MATCHED EXAMPLE 10 | Finding the variance of a continuous probability density function

The probability density function for a continuous random variable X is given by

$$f(x) = \begin{cases} \dfrac{5}{32}x^4 & \text{if } 0 \le x \le 2 \\ 0 & \text{elsewhere} \end{cases}$$

Find the variance.

Steps	Working
1 Write the formula for the variance and calculate $E(X)$ and $E(X^2)$.	
μ or $E(X)$ was calculated in Matched example 7.	
$E(X^2) = \displaystyle\int_{-\infty}^{\infty} x^2 f(x)\, dx$	
2 Substitute into the variance formula.	

MATCHED EXAMPLE 11	Using the symmetry properties of the standard normal distribution.

A continuous random variable Z is normally distributed with a mean of 0 and a standard deviation of 1. If $\Pr(Z \leq 1.71) = 0.9564$, find $\Pr(-1.71 \leq Z \leq 1.71)$.

Steps	Working
1 Draw normal distribution curves that illustrate $\Pr(Z \leq 1.71) = 0.9564$ and $\Pr(-1.71 \leq Z \leq 1.71)$.	
2 To find $\Pr(-1.71 \leq Z \leq 1.71)$ (the region shaded under the second curve), first find $\Pr(Z \geq 1.71)$ (the region shaded under the first curve).	
3 By symmetry, subtract double the value found above from 1.	

MATCHED EXAMPLE 12 | Finding probabilities using the formula for *z*

SB

p. 511

A continuous random variable X is normally distributed with a mean of 60 and a standard deviation of 12. If $\Pr(Z \le 2) = 0.9772$, find $\Pr(X < 36)$.

Steps	Working
1 Draw a normal distribution curve that illustrates $\Pr(Z \le 2) = 0.9772$.	
2 Find z, the standard normal value for $x = 36$.	
3 Find $\Pr(X < 36)$ using the equivalent area from a standard normal curve. Draw the required area on a normal curve and calculate by symmetry.	

SB

Using CAS 5:
Probabilities
for a normally
distributed random
variable
p. 512

MATCHED EXAMPLE 13	Applying the normal distribution

Year 5 students complete a strength test. The times taken to complete the test are normally distributed with a mean of 20 minutes and a standard deviation of 5 minutes.

a Find, correct to four decimal places, the proportion of students who complete the test in less than 18 minutes.

b Students who complete the task in a time between 15 minutes and 25 minutes are classified as having average fitness levels. Find, correct to four decimal places, the probability of a student selected at random being classified as having average fitness.

c Find, correct to four decimal places, the probability of a student with average fitness completing the task in less than 18 minutes.

Steps	Working

a 1 Draw the normal distribution curve, and show the three standard deviations above and below the mean on the horizontal axis scale.

2 Calculate $\Pr(X < 18)$ using CAS.

TI-Nspire	ClassPad

b Calculate $\Pr(15 < X < 25)$.

c 1 Write the conditional probability formula.

The probability of a student completing a task in under 18 minutes given they have average fitness is

$$\Pr(A|B) = \frac{\Pr(A \cap B)}{\Pr(B)}$$

$$(X < 18) \cap (15 < X < 25)$$

$$= (15 < X < 18)$$

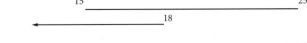

SB

Using CAS 6:
Inverse cumulative
normal distribution
p. 514

2 Calculate using CAS.

MATCHED EXAMPLE 14 | Applying the inverse cumulative normal distribution

A continuous random variable X is normally distributed with a mean of 35. If $\Pr(X \geq 25) = 0.65$, find the standard deviation correct to two decimal places.

Steps	Working
1 Draw the normal distribution curve for X and the corresponding normal distribution curve for Z.	
2 Use the CAS inverse normal distribution to find c. **TI-Nspire** **ClassPad**	
3 Substitute into $z = \dfrac{x - \mu}{\sigma}$ and solve for σ, algebraically or using CAS.	

CHAPTER

12 SAMPLE PROPORTIONS

p. 533

MATCHED EXAMPLE 1	Parameters and statistics

The first 30 people leaving the Chadstone Shopping Centre after 8.30 pm were asked how much they spent. The smallest amount was $25.50, the largest was $120.50 and the average amount was $55.50. Identify the population, some parameters and statistics.

Steps	Working
1 The population is the whole group being investigated.	
2 Parameters are values for the whole population.	
3 Statistics are values from the sample.	

MATCHED EXAMPLE 2 | Sample proportion 1

A college has 2050 students, of whom 875 do not study mathematics.

What proportion of the students study mathematics? Write the answer correct to three decimal places.

Steps	Working
Simplify the fraction if possible.	

MATCHED EXAMPLE 3 | Sample proportion 2

From a sample of 50 people, 15 always had wine at their home. What is the sample proportion of people having wine at home?

Steps	Working
1 Calculate the proportion, with $x = 15$ and $n = 50$.	
2 Write the answer.	

MATCHED EXAMPLE 4 | Sample proportion probability

About 55% of Australians own a house. What is the probability, correct to three significant figures, that a random sample of 12 Australians will have

a exactly 5 people who own a house?

b fewer than 2 people who own a house?

Steps	Working
a 1 Find the sample proportion.	
2 Use the binomial probability.	
3 Do the calculation.	
4 Write the answer.	
b 1 Write the probability as a sum.	
2 Use the binomial probability.	
3 Do the calculations.	
4 Write the answer.	

12

p. 541

MATCHED EXAMPLE 5 | Expected sample proportion

The probability of drawing an ace from a biased deck of cards is $\dfrac{7}{26}$. A card is drawn 30 times, and the proportion of drawing an ace is calculated. This sampling of 30 is repeated many times. What is the expected sample proportion of drawing an ace?

Steps	Working
1 Use the rule that $E(\hat{p}) = p$.	
2 Write the answer.	

SB

p. 541

Seventy samples of 30 people each are chosen at random to check whether they are artists. A total of 300 people are artists. Estimate the probability of the people not being artists, correct to two significant figures.

Steps	Working
1 Find the mean sample proportion.	
2 Write the answer.	

SB

Using CAS 3:
Simulating
sample proportion
distributions
p. 544

SB

p. 546

MATCHED EXAMPLE 7 | Sample proportion standard deviation

A deck of cards is loaded so that the probability of drawing an ace is $\dfrac{7}{26}$. A card is drawn 30 times, and the proportion of drawing an ace is calculated. This process is repeated many times. What is the standard deviation of the proportion of drawing an ace?

Steps	Working
1 Use the rule that $\mathrm{SD}(\hat{p}) = \sqrt{\dfrac{p(1-p)}{n}}$.	
2 Write the answer.	

MATCHED EXAMPLE 8	Normal approximation for sample proportion

Given that about 10% of Australians are artists, what is the probability that in a sample of 300 Australians 30 to 40 of them are artists?

Steps	Working
1 Check the values of np and nq.	
2 State the conclusion.	
3 Find the mean and standard deviation.	
4 Find the sample proportions required for the integer interval $[30, 40]$. Use the cumulative normal distribution. **TI-Nspire**　　　　　　　　　**ClassPad**	
5 Write the answer.	

12

MATCHED EXAMPLE 9	Standard error estimation

Eighteen out of 45 randomly selected people said that the VFA club they most liked was Richmond. Estimate the proportion of people who like Richmond the most, and estimate the standard deviation of the sampling distribution.

Steps	Working
1 Find the value of $\hat{p}$.	
2 Use it as an estimate of p.	
3 Estimate the standard error.	
4 State the answer.	

MATCHED EXAMPLE 10 | Sample size required

A study found the sample proportion of people with an adverse reaction to a new anti-allergen was about 0.05. What sample size would be needed to ensure that the standard deviation of the sampling distribution was less than 0.003?

Steps	Working
1 Estimate the value of p.	
2 Write the formula for SD.	
3 Substitute and solve for n.	
4 Write the answer.	

SB

Using CAS 4:
Standard normal
quantiles and
confidence
intervals
p. 551

MATCHED EXAMPLE 11 | 99% confidence interval

A random sample of 60 items from a large population gave a sample proportion of $\hat{p} = 0.45$ for a particular property. What is the 99% confidence interval for the probability of the property, correct to three decimal places?

Steps	Working
1 Check that the normal distribution can be used.	
2 Check the number of successes.	
3 Find the proportion below the interval.	
4 **TI-Nspire** **ClassPad**	
Use the **invNorm** function to find the quantiles. Use the **invNormCDf** function to find the quantiles.	
Alternatively, find the **zInterval**. Note that the margin of error is also displayed. Alternatively, find the **One-sample Z Int.**	
5 State the margin of error.	
6 Estimate σ and simplify.	
7 Find the ends of the confidence interval.	
8 Write the answer.	

MATCHED EXAMPLE 12 | 95% confidence interval

Two hundred pumpkin seeds were moistened and placed in an incubator. When they were checked nine days later, 150 were found to have germinated. Estimate the 95% confidence interval for the germination rate, correct to three decimal places.

Steps	Working
1 Find the sample proportion.	
2 State the value of $z_{0.95}$.	
3 Use the formula for the confidence interval.	
4 Substitute in the values.	
5 Calculate the answer.	
6 State the confidence interval.	

TI-Nspire

ClassPad

1 Open the **Statistics** application.

2 Tap **Calc** > **Interval**.

1 Press **menu** > **Statistics** > **Confidence Intervals** > **1–Prop z Interval**.

2 Enter the values as shown above.

3 Press **enter**

3 In the lower window, select **One-Prop Z Int** from the drop-down menu then tap **Next**.

4 Enter the value as shown above then tap **Next**.

4 The z-interval table will be displayed.

5 The **CLower** and **CUpper** values of the confidence interval are highlighted above.

5 The values will be displayed in the lower window.

6 The **Lower** and **Upper** values of the confidence interval are highlighted above.

MATCHED EXAMPLE 13 | Sample size for a confidence interval

About 50% of Year 12 students obtain a part-time job before they complete Year 12. How large a sample would be needed to establish a 90% confidence level within a margin of error of 5%?

Steps	Working
1 Write the formula for the margin of error.	
2 Write the known values.	
3 Substitute.	
4 Solve for n.	
5 Write the answer.	

9780170464062

SB

p. 555

A survey of school buses found that the 95% confidence interval for the proportion that had at least one with a compartmentalisation feature was about $(0.12, 0.42)$. How many vans were surveyed?

Steps	Working
1 $\hat{p}$ is the centre of the confidence interval.	
2 The margin of error is half the confidence interval.	
3 Write the formula.	
4 Substitute values and solve for n.	
5 Write the answer.	

12

MATCHED EXAMPLE 15 | Sample size needed for percentage confidence interval

A survey is done to establish the percentage of Year 12 students from co-parent families to within 7% at a confidence level of 90%. What is the minimum number to survey?

Steps	Working
1 Assume the worst case, $\hat{p}$.	
2 Write the margin of error.	
3 Write the formula.	
4 Substitute values and solve for n.	
5 Write the answer.	

9780170464062

Answers

Worked solutions available on Nelson MindTap.

CHAPTER 1

MATCHED EXAMPLE 1

a The given expression is not a polynomial.

b The given expression is a polynomial.

Degree = 4

Leading term $= -3^{\frac{1}{2}} x^4$

Coefficients $= -3^{\frac{1}{2}}, \left(\dfrac{1}{3}\right)^3, \dfrac{1}{2}$

MATCHED EXAMPLE 2

a $2x^3 + 5x - 1$

Degree 3

b $6x^5 + 23x^4 + 48x^3 + 34x^2 + 15x$

Degree 5

MATCHED EXAMPLE 3

$m = 1$

MATCHED EXAMPLE 4

a $h(t) = -0.125t^2 + 6t - 40$

b 7.5 cm

c The plant height will be 0 cm at 8°C and 40°C.

MATCHED EXAMPLE 5

a $2M(x) - 4N(x) = (2x^4 + 4x^3 + 2qx^2 + 6x + 2p) - (8x^2 - 4qx + 12)$
$= 2x^4 + 4x^3 + 2qx^2 + 6x + 2p - 8x^2 + 4qx - 12$
$= 2x^4 + 4x^3 + (2q - 8)x^2 + (6 + 4q)x + 2p - 12$

b $q = 4, p = 2$

MATCHED EXAMPLE 6

$x^3 - 10x^2 + 30x + 6 = (x - 5)(x^2 - 5x + 5) + 31$

MATCHED EXAMPLE 7

$2x + 3 + \dfrac{1}{x+1}$

MATCHED EXAMPLE 8

$Q(x) = -x + 6$, remainder $= 10x + 18$

MATCHED EXAMPLE 9

The remainder is $4\dfrac{3}{4}$.

MATCHED EXAMPLE 10

$k = 5$

MATCHED EXAMPLE 11

a $f(1) = 2(1)^3 - a(1)^2 + b(1) + 3$

$3 = -a + b + 5$

$a - b = 2$

b $f(y) = 2y^3 - 4y^2 + 2y + 3$

MATCHED EXAMPLE 12

$x^3 - 5x^2 + 2x + 8 = (x + 1)(x - 2)(x - 4)$

MATCHED EXAMPLE 13

$f(x) = (x - 2)(x + 2)(x + 3)(x + 2)$

MATCHED EXAMPLE 14

$f(x) = (x - 1)(x - 2)(2x + 1)$

MATCHED EXAMPLE 15

a $(3y - 1)(9y^2 + 3y + 1)$

b $(x + 5)(x^2 - 2x + 13)$

MATCHED EXAMPLE 16

$p = 1$

MATCHED EXAMPLE 17

$x = 0$ or $x = 1$ or $x = -\dfrac{1}{2}$ or $x = -1$

MATCHED EXAMPLE 18

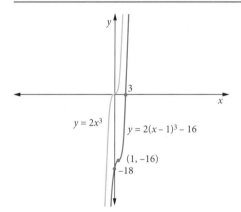

MATCHED EXAMPLE 19

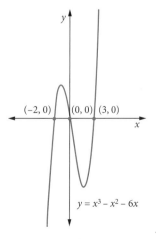

MATCHED EXAMPLE 20

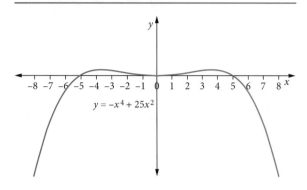

$y = -x^4 + 25x^2$

MATCHED EXAMPLE 21

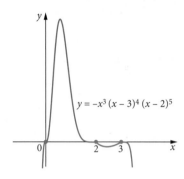

$y = -x^3(x-3)^4(x-2)^5$

MATCHED EXAMPLE 22

$y = -4(x-2)^3 + 3$

MATCHED EXAMPLE 23

a Maximum profit = \$13 000 and no. of toys = 26.

b Zeros: $x = -20, 6, 40$

c $P(x) = -(x+20)(x-6)(x-40)$

d $x \in (6, 40)$

CHAPTER 2

MATCHED EXAMPLE 1

$\dfrac{-4 \pm 2\sqrt{7}}{3}$

MATCHED EXAMPLE 2

$k \in (4, 8)$

MATCHED EXAMPLE 3

$k = 2$ and $k = 6$ produces no solution.

MATCHED EXAMPLE 4

The equation will have one solution for $m Î R \setminus \{1, 4\}$.

MATCHED EXAMPLE 5

$x = \left\{ \dfrac{\pi}{3}, \dfrac{4\pi}{3} \right\}$

MATCHED EXAMPLE 6

$x = \left\{ \dfrac{3\pi}{8}, \dfrac{5\pi}{8}, \dfrac{11\pi}{8}, \dfrac{13\pi}{8} \right\}$

MATCHED EXAMPLE 7

$x = \left\{ \dfrac{\pi}{12}, \dfrac{7\pi}{12}, \dfrac{13\pi}{12}, \dfrac{19\pi}{12} \right\}$

MATCHED EXAMPLE 8

$x = -\dfrac{\pi}{6} + \dfrac{1}{2}n\pi, \quad x = \dfrac{\pi}{6} + \dfrac{1}{2}n\pi$

MATCHED EXAMPLE 9

$x = \dfrac{\pi}{3} + n\pi, \quad n \in Z$

MATCHED EXAMPLE 10

$x = 5$

MATCHED EXAMPLE 11

$x = 1, 2$

MATCHED EXAMPLE 12

$x = \log_e(2)$

MATCHED EXAMPLE 13

$x = 2$

MATCHED EXAMPLE 14

$x = 4$

MATCHED EXAMPLE 15

$y = 2 - \dfrac{3q}{p}$

MATCHED EXAMPLE 16

$x = -\dfrac{1}{4} \log_a (2c - 3ky)$

MATCHED EXAMPLE 17

$x = ya^{\frac{k+c}{2}} - b$

MATCHED EXAMPLE 18

$x_1 = -2$

MATCHED EXAMPLE 19

$x_2 \approx -0.4545$

CHAPTER 3

MATCHED EXAMPLE 1

The maximal domain is $R \setminus \{2, 3\}$.

MATCHED EXAMPLE 2

The range of f: $[2,5) \to R$, $g(x) = -x^2 + 6x - 5$ is $[-3, 0)$.

MATCHED EXAMPLE 3

The range of the function is $[-81, \infty)$.

MATCHED EXAMPLE 4

f and h are one-to-one, but g is not.

MATCHED EXAMPLE 5

$(-3, 4)$

MATCHED EXAMPLE 6

5 units

MATCHED EXAMPLE 7

2

MATCHED EXAMPLE 8

$2x - 5y + 18 = 0$

MATCHED EXAMPLE 9

$4x + 3y - 8 = 0$

MATCHED EXAMPLE 10

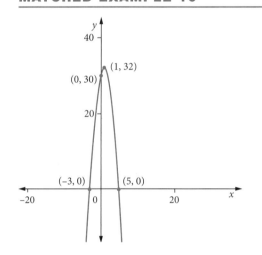

MATCHED EXAMPLE 11

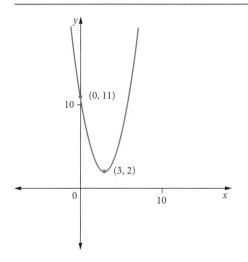

MATCHED EXAMPLE 12

There are no intersections.

MATCHED EXAMPLE 13

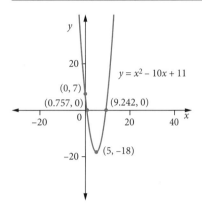

MATCHED EXAMPLE 14

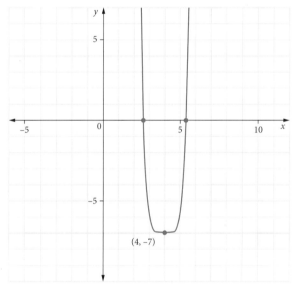

The domain is R, the real numbers.

The range is $y \geq -7$.

MATCHED EXAMPLE 15

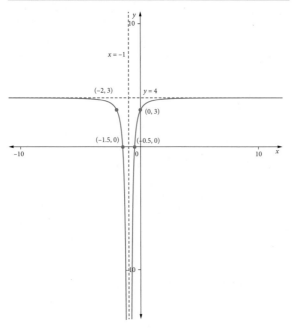

The domain is $\{x \in R: x \neq -1\}$.

The range is $\{x \in R: x < -4\}$.

MATCHED EXAMPLE 16

a

$$y = \frac{2x+1}{x+1}$$

$$= \frac{2(x+1)-1}{x+1}$$

$$= \frac{2(x+1)}{x+1} - \frac{1}{x+1}$$

$$= 2 - \frac{1}{x+1}$$

b

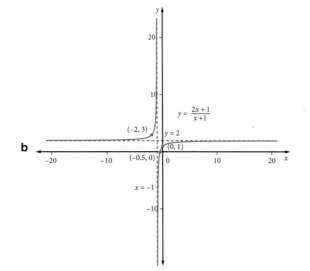

The domain is $R\backslash\{-1\}$ and the range is $R\backslash\{2\}$.

MATCHED EXAMPLE 17

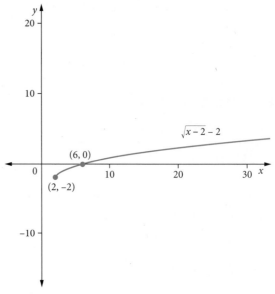

The domain is $D = \{x \in R: x \geq 2\}$.

The range is $\{x \in R: x \geq -2\}$.

MATCHED EXAMPLE 18

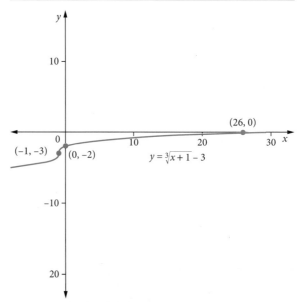

The domain and range of the function $y: R \to R$, $y = \sqrt[3]{x+1} - 3$ are all real numbers.

MATCHED EXAMPLE 19

No

MATCHED EXAMPLE 20

$f^{-1}: R \to R, f^{-1}(x) = \sqrt[3]{\dfrac{x+5}{2}}$

MATCHED EXAMPLE 21

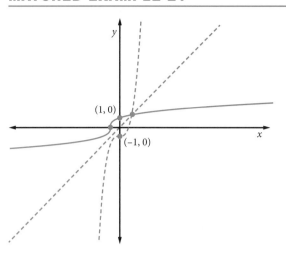

MATCHED EXAMPLE 22

$x = \log_{10}(5y - 2) + 3$

MATCHED EXAMPLE 23

$x = \dfrac{e^{7-y^3} - 7}{5}$

MATCHED EXAMPLE 24

$\log_3(2)$

MATCHED EXAMPLE 25

The inverse points are $(-1.8, -1)$, $(0, -0.28)$ and $(3,0)$. The inverses of the other 3 points are $(4.9, 0.1)$, $(7.5, 0.2)$ and $(11, 0.3)$.

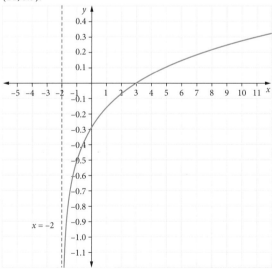

MATCHED EXAMPLE 26

$\dfrac{7\pi}{18}$

MATCHED EXAMPLE 27

$-\dfrac{1}{\sqrt{2}}$

MATCHED EXAMPLE 28

$\cot(x) = -\dfrac{4}{3}$.

MATCHED EXAMPLE 29

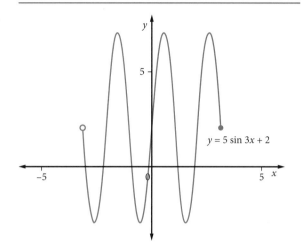

MATCHED EXAMPLE 30

1 $T = 5$

2 amplitude $= 3$

3 central value $= 1$

MATCHED EXAMPLE 31

The equation could be $f(x) = 4\sin(8\pi x) + 2$.

CHAPTER 4

MATCHED EXAMPLE 1

The average rate of change between $x = 3$ and $x = 10$ is 26.

MATCHED EXAMPLE 2

The value of the derivative is $f'(3) = 108$.

MATCHED EXAMPLE 3

A Since $f(x)$ is not continuous, it cannot be differentiable.

B $f(x)$ is not continuous at $x = 3$, so it is not differentiable at $x = 3$.

C $f(x)$ is continuous but not smooth at $x = 2$, so it is not differentiable.

MATCHED EXAMPLE 4

a $f'(x) = 6x^2 + \dfrac{33\sqrt{5}}{4} x^{\frac{7}{4}}$

b $f'(x) = 18x^8$

c $f'(x) = -\dfrac{25}{28x^{\frac{9}{4}}}$

MATCHED EXAMPLE 5

a $f'(1) = 44$

b $f'(2) = -1$

c $f'\left(\dfrac{2}{3}\right) = 9$

MATCHED EXAMPLE 6

$f'(x) = 42x^6 - 165x^4 + 6x^2 - 11$

MATCHED EXAMPLE 7

$f'(-1) = 0$

MATCHED EXAMPLE 8

$\dfrac{d}{dx}\left(\dfrac{5x+2}{7x-1}\right) = -\dfrac{19}{(7x-1)^2}$

MATCHED EXAMPLE 9

a $\dfrac{d}{dx}\left[\dfrac{1}{(7x+5)^3}\right] = -\dfrac{21}{(7x+5)^4}$

b $\dfrac{d}{dx}\left(\sqrt{3x-1}\right) = \dfrac{3}{2\sqrt{3x-1}}$

MATCHED EXAMPLE 10

$12x^3(2x+5)(3x+5)$

CHAPTER 5

MATCHED EXAMPLE 1

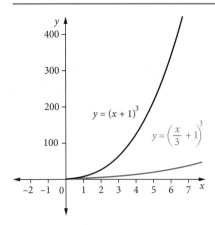

For each point on $y = (x+1)^3$, the distance from the y-axis is three times on the graph of $y = \left(\dfrac{x}{3}+1\right)^3$.

The graph is dilated from the y-axis by a factor 3.

MATCHED EXAMPLE 2

1

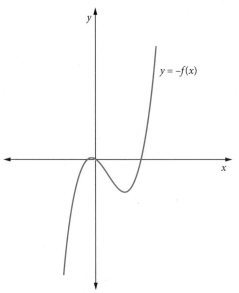

2

MATCHED EXAMPLE 3

1

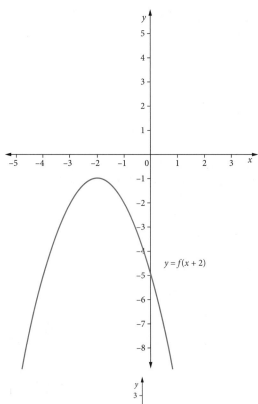

2

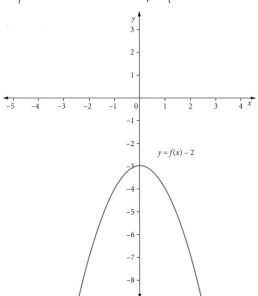

MATCHED EXAMPLE 4

Q is the point $(-3, -2)$.

MATCHED EXAMPLE 5

The inverse transformation $g(x) \rightarrow \left[g\left(-\dfrac{x}{3} - 2 \right) + 4 \right]$ is a

translation of 4 units up, a dilation parallel to the x direction by a factor of 3, a reflection in the y-axis, and a translation of 2 units in the positive x direction.

MATCHED EXAMPLE 6

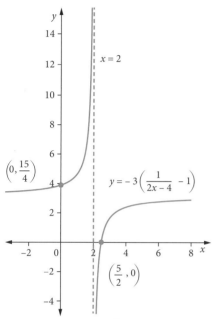

The rule is $f(x) = 3 - \dfrac{3}{2x-4}$, with domain $R - \{2\} \rightarrow R$ and range R.

MATCHED EXAMPLE 7

a $f(x)$ has been dilated from the x-axis by a factor $\dfrac{1}{3}$,

reflected in the x-axis, translated 2 units in the positive x direction and translated 5 units in the positive y direction.

b f is reflected in the y-axis and reflected in the x-axis, translated 3 units to the left and translated 4 units down.

c f is dilated from the x-axis by a factor 3, dilated from the y-axis by a factor 2, reflected in the y-axis, translated 2 units to the left and translated 4 units down.

MATCHED EXAMPLE 8

$f(x) = \dfrac{1}{4(x-1)^2} - 4$

The graph of $y = \dfrac{1}{x^2}$ is dilated by a factor of $\dfrac{1}{4}$ from the y-axis and translated 1 unit to the right and 4 units down.

MATCHED EXAMPLE 9

E: $f(x) = -2(x+1)^2 + 3$ is correct.

MATCHED EXAMPLE 10

a $\begin{bmatrix} -1 & 0 \\ 0 & 4 \end{bmatrix} \times \begin{bmatrix} x \\ y \end{bmatrix} + \begin{bmatrix} 1 \\ 3 \end{bmatrix}$

b $\begin{bmatrix} \dfrac{1}{3} & 0 \\ 0 & -1 \end{bmatrix} \times \begin{bmatrix} x \\ y \end{bmatrix} + \begin{bmatrix} 6 \\ 5 \end{bmatrix}$

MATCHED EXAMPLE 11

The image is given by $y = -\dfrac{1}{2}\cos\left(\dfrac{x}{\pi} + \dfrac{1}{4} \right) - 1$.

MATCHED EXAMPLE 12

$f(x) = \dfrac{x^2}{2} - \dfrac{11x}{2} + 15$

MATCHED EXAMPLE 13

a Maximal domain $= [-5, 5] \, Ç \, (-\infty, 0) = [-5, 0)$

b Maximal domain of inverse $= [-14\,640, 0)$

MATCHED EXAMPLE 14

The functions could be $y_1 = \log_3 4x$ and $y_2 = \log_3 (x+2)$.

MATCHED EXAMPLE 15

a
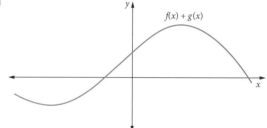

b
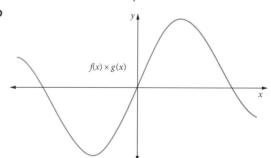

MATCHED EXAMPLE 16

a 2

b 3

c −2

d −2

MATCHED EXAMPLE 17

a $f \circ g(1.5) = 12$

b $\dfrac{9}{x^2} + \dfrac{6}{x} + 4$

c Range $= (-\infty, 0) \cup (0, \infty)$

d $h(0.25) = 172$

MATCHED EXAMPLE 18

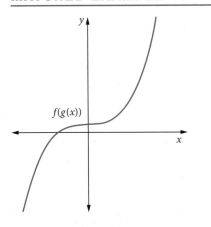

MATCHED EXAMPLE 19

a $f(g(x)) = x^2 + 5x + 5$ has domain R and range $\left[-\dfrac{5}{4}, \infty\right)$.

b $g(f(x)) = x^2 + x + 1$ has domain R and range $\left[\dfrac{3}{4}, \infty\right)$.

c $a = 1, b = 1, c = 2$ and $d = 2$

MATCHED EXAMPLE 20

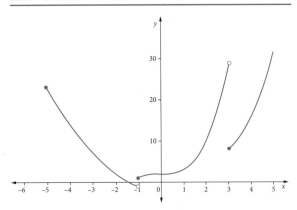

MATCHED EXAMPLE 21

1 $p(x) = \begin{cases} 2.56x + 6 & \text{for } 0 < x \le 12 \\ 1.92x & \text{for } x > 12 \end{cases}$

2 $23.92

3 $46.08

CHAPTER 6

MATCHED EXAMPLE 1

a $x < -2, -1 < x < \dfrac{2}{5}, x > 1$

or in interval notation:

$(-\infty, -2) \cup (-1, \dfrac{2}{5}) \cup (1, \infty)$

b $-2 < x < -1, \dfrac{2}{5} < x < 1$

or in interval notation:

$(-2, -1) \cup (\dfrac{2}{5}, 1)$

MATCHED EXAMPLE 2

a $x \in (-\infty, -1] \cup [1, \infty)$

b $x \in [-1, 1]$

MATCHED EXAMPLE 3

Coordinates are $(-\dfrac{4}{5}, \dfrac{16}{25})$, which is a local max, and $(0, 0)$, which is a local min.

MATCHED EXAMPLE 4

The stationary point of inflection is at $(2, 0)$.

MATCHED EXAMPLE 5

Absolute maximum $(0.667, 4.741)$

MATCHED EXAMPLE 6

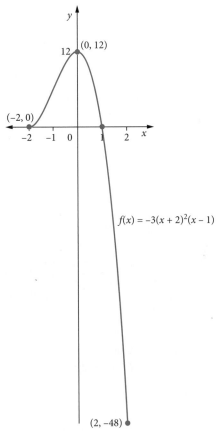

Range: $[-48, 12]$

MATCHED EXAMPLE 7

a The tangent meets the graph again at $(-3, -16)$.

b

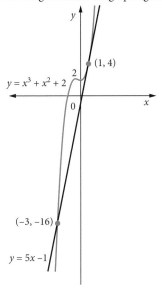

MATCHED EXAMPLE 8

The maximum possible volume of cone is $9.9\,\text{cm}^3$.

MATCHED EXAMPLE 9

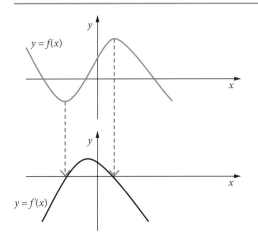

MATCHED EXAMPLE 10

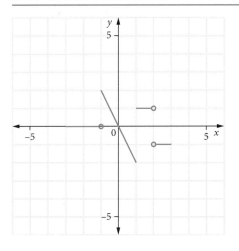

MATCHED EXAMPLE 1

$$x^4 - \frac{x^2}{2} - 7x + c$$

MATCHED EXAMPLE 2

$$\int (7x+3)^5 \, dx = \frac{1}{42}(7x+3)^6 + c$$

MATCHED EXAMPLE 3

$$y = x^4 + x^2 - x - 87$$

MATCHED EXAMPLE 4

A $v = 2t^2 + 2t - 4$

B $x = \dfrac{2t^3}{3} + t^2 - 4t$

MATCHED EXAMPLE 5

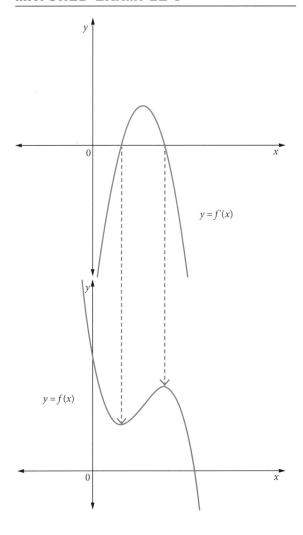

MATCHED EXAMPLE 6

The area is approximately $6.5\,\text{units}^2$.

MATCHED EXAMPLE 7

a Total area of left rectangles $= \dfrac{55}{2}$ units^2.

b Total area of right rectangles $= \dfrac{85}{4}$ units^2.

MATCHED EXAMPLE 8

A $\dfrac{7}{4}$

B $\dfrac{202}{3}$

MATCHED EXAMPLE 9

$\int_{3}^{1}\left(g(x)+6\right)dx = -14$

MATCHED EXAMPLE 10

$\int_{0}^{1}2\left(f(x)+3\right)dx = 14$

MATCHED EXAMPLE 11

$\dfrac{158}{3}$ units^2

MATCHED EXAMPLE 12

Area $= \dfrac{38}{9}$ units^2

MATCHED EXAMPLE 13

Area $= \dfrac{9}{2}$ units^2

MATCHED EXAMPLE 14

16 metres

MATCHED EXAMPLE 15

$\dfrac{1}{6}$

CHAPTER 8

MATCHED EXAMPLE 1

The minimum is -1 at $x = 2$, $x = 0$.

The maximum is 7 at $x = 1$, $x = 3$.

MATCHED EXAMPLE 2

a $a = -2$: (vertical) dilation from the x-axis by a factor of 2, amplitude $= 2$

$a < 0$: *(reflection)* reflected in the x-axis

$b = \pi$: (horizontal) translation π units left

$c = -1$: (vertical) translation 1 unit down

$n = 3$: (horizontal) dilation from the y-axis by a factor of $\dfrac{1}{3}$, period $= \dfrac{2\pi}{3}$

b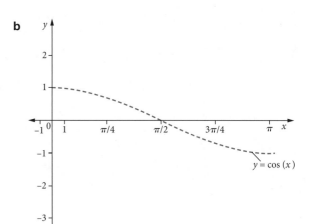

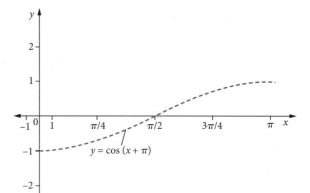

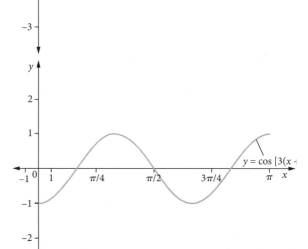

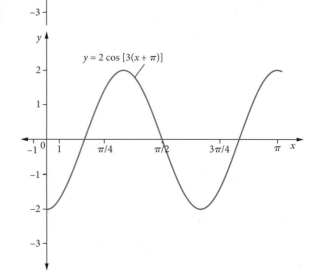

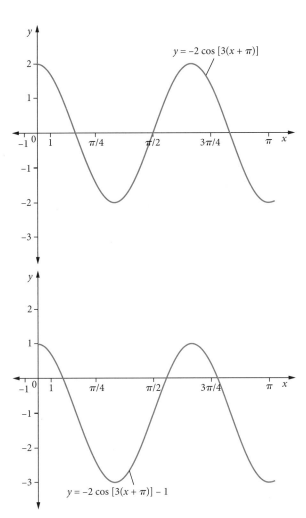

$y = -2 \cos [3(x + \pi)]$

$y = -2 \cos [3(x + \pi)] - 1$

c The range is $[-3, 1]$.

MATCHED EXAMPLE 3

a $g(x) = -\dfrac{1}{2} \tan(5x) + 3$

b The graph of $f(x)$ is dilated from the y-axis by a factor of $\dfrac{1}{2}$ and reflected in the x-axis. The graph of $f(x)$ is dilated from the y-axis by a factor of $\dfrac{1}{5}$. The graph of $f(x)$ is translated up by 3 units.

MATCHED EXAMPLE 4

The derivative of $y = \tan(x) \cos(x)$ is $\dfrac{1 - \sin^2(x)}{\cos(x)}$.

MATCHED EXAMPLE 5

$\dfrac{-x^3 \sin(x) + \sin(x) - 3x^2 \cos(x)}{(x^3 - 1)^2}$

MATCHED EXAMPLE 6

$-\sin(\sin(x))\cos(x)$

MATCHED EXAMPLE 7

$\dfrac{4x^3 \cos(x^2) + 2x^5 \sin(x^2)}{\cos^2(x^2)}$

MATCHED EXAMPLE 8

$y - 1 = 0$

MATCHED EXAMPLE 9

a $v = -\sin(2t)$

b 0 cm/s

c $t = 0$ s

d The maximum acceleration is 2 cm/s^2 when $\cos(2t) = -1$.

MATCHED EXAMPLE 10

$t_0 = 5.8$, $t_1 \approx 6.32466$, $t_2 \approx 6.28316$

The object takes 6.283 seconds.

MATCHED EXAMPLE 11

a $F(x) = 4\tan(2x + 1) - \dfrac{1}{5}\cos(5x) + c$

b $F(x) = \dfrac{1}{2}\tan(2x) + \dfrac{1}{5}\cos(5x) + c.$

MATCHED EXAMPLE 12

$\displaystyle\int \left[3(x^2 + 1)\cos(x) + 6x\sin(x) \right] dx = 3(x^2 + 1)\sin x + c$

MATCHED EXAMPLE 13

$f(x) = \tan(2x) - \dfrac{9}{8}$

MATCHED EXAMPLE 14

a $\dfrac{\sqrt{3}}{2}$

b 0

MATCHED EXAMPLE 15

Area $= 2$ units^2

MATCHED EXAMPLE 16

a

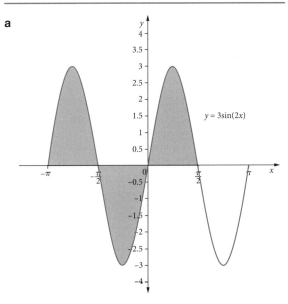

$y = 3\sin(2x)$

b The shaded area is 9 units^2.

CHAPTER 9

MATCHED EXAMPLE 1

a Sketch the graph of $f(x) = e^{2x}$

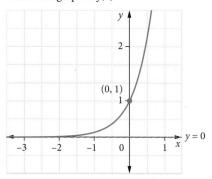

b

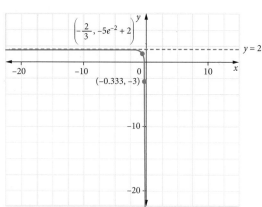

MATCHED EXAMPLE 2

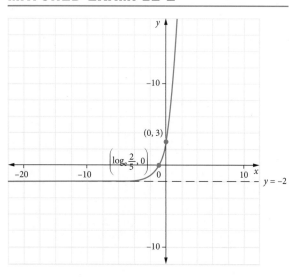

MATCHED EXAMPLE 3

a $b = 1$
$a = 6$

b $y = 6e^x + 1$

MATCHED EXAMPLE 4

$y = \dfrac{5}{9}(3)^t$

MATCHED EXAMPLE 5

It takes about 54 minutes for the hot chocolate to become undrinkable.

MATCHED EXAMPLE 6

a There were 500 trees at the beginning.

b There were about 710 trees after 7 years.

MATCHED EXAMPLE 7

a $\dfrac{dy}{dx} = 7e^{7x+1}$

b $\dfrac{dy}{dx} = (4x - 1)e^{2x^2 - x + 3}$

MATCHED EXAMPLE 8

a $\dfrac{dy}{dx} = -6e^{3x}\sin(2e^{3x})$

b $x^2 e^{-5x}(-5x + 3)$

MATCHED EXAMPLE 9

$\dfrac{37e^{\frac{5}{2}}}{54}$

MATCHED EXAMPLE 10

$y = 9.986x - 5.844$

$y = 0.100x + 14.328$ (correct to three decimal places)

MATCHED EXAMPLE 11

a $T_0 = 900°C$.

b The temperature difference is about 6.1°C.

c The temperature after 10 minutes is about 24.1°C.

d After 10 minutes, the metal is cooling at about 3.0°C/minute.

MATCHED EXAMPLE 12

a 70.150 mg of Radon-226 remains.

b Radon-226 is decaying at a rate of 2.715 mg/day.

MATCHED EXAMPLE 13

a $\displaystyle\int e^{5-x}\,dx = -e^{5-x} + c$

b $\displaystyle\int \dfrac{e^{2x} + 7 + e^{6x}}{3e^{5x}}\,dx = -\dfrac{1}{9}e^{-3x} - \dfrac{7}{5}e^{-5x} + e^x + c$

MATCHED EXAMPLE 14

$f(x) = \dfrac{2}{5}e^{5x} + \dfrac{13}{5}$

MATCHED EXAMPLE 15

3.15

MATCHED EXAMPLE 16

a

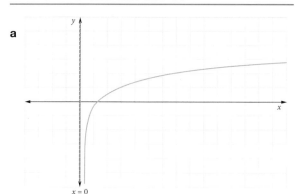

b

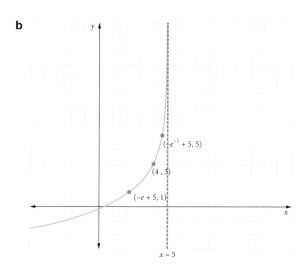

MATCHED EXAMPLE 17

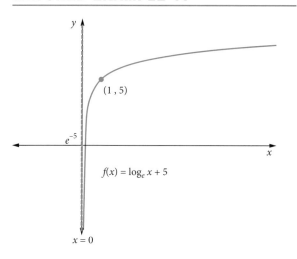

MATCHED EXAMPLE 18

$y = 2\ln(x + 5)$

MATCHED EXAMPLE 19

a The derivative of $\log_e(7x + 1)$ is $\dfrac{7}{7x+1}$.

b $\dfrac{dy}{dx} = \dfrac{6}{3x-2}$

c $\dfrac{dy}{dx} = \dfrac{3}{2x^2 \log_e\left(2x^3\right)^{\frac{1}{2}}}$

d $\dfrac{dy}{dx} = \dfrac{1}{(x-1)\log_e(5)}$

MATCHED EXAMPLE 20

$2x^2 + 6x^2 \ln(5x)$

MATCHED EXAMPLE 21

a $y = 2x + \log_e 3 - 2$

b $y = \left(\dfrac{2}{a}\right)x + \log_e(3a^2) - 2$

MATCHED EXAMPLE 22

$\displaystyle\int \dfrac{3}{5-2x}\,dx = -\dfrac{3}{2}\log_e(5-2x) + c \text{ for } x < \dfrac{5}{2}$

MATCHED EXAMPLE 23

a $\dfrac{1}{5}\log_e\left(\dfrac{7}{2}\right)$

b $4 - 4\log_e\left(\dfrac{5}{2}\right)$

MATCHED EXAMPLE 24

$f(x) = -\dfrac{3}{2}\log_e(7-2x) + \dfrac{3}{2}\log_e 3$

MATCHED EXAMPLE 25

$\displaystyle\int 2x\log_e(x)\,dx = x^2\log_e x - \dfrac{1}{2}x^2 + c$

MATCHED EXAMPLE 26

$k = \dfrac{6e^9 - 1}{5}$

MATCHED EXAMPLE 27

a $\displaystyle\int \dfrac{3x^2+1}{x^3+x-2}\,dx = \log_e(x^3 + x - 2) + c$

b $\displaystyle\int \dfrac{16x}{4x^2-3}\,dx = 2\ln(4x^2 - 3) + c$

MATCHED EXAMPLE 28

$\dfrac{1}{4}\left[\log_e\left(\dfrac{13}{3}\right)\right]$

MATCHED EXAMPLE 29

$1 - \dfrac{2\log_e(2)}{5}$

MATCHED EXAMPLE 30

a It will take Amy about 76 days to learn the very basic vocabulary.

b $\dfrac{dw}{dt} = \dfrac{90}{t-1}$

c Amy would be learning one word a day after 91 days.

CHAPTER 10

MATCHED EXAMPLE 1

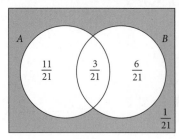

$$\Pr(A' \cap B') = \frac{1}{21}$$

MATCHED EXAMPLE 2

$$\Pr(A \cap B') = 0.12$$

MATCHED EXAMPLE 3

$p = 0.3$ (p cannot be negative.)

MATCHED EXAMPLE 4

$$\Pr(1 \text{ blue-coloured}) = \frac{8}{17}$$

MATCHED EXAMPLE 5

$$\Pr(CC) = 0.06$$

MATCHED EXAMPLE 6

$$\Pr(B) = \frac{12}{35}$$

MATCHED EXAMPLE 7

$$\Pr(\text{jeep and car}|\text{two start}) = \frac{1}{13}$$

MATCHED EXAMPLE 8

The probability distribution of X is

x	0	1	2
$\Pr(X = x)$	$\frac{1}{7}$	$\frac{18}{35}$	$\frac{12}{35}$

MATCHED EXAMPLE 9

$$p = \frac{2}{3}$$

MATCHED EXAMPLE 10

$$\Pr(\text{same on Fri and Sat}) = 0.42$$

MATCHED EXAMPLE 11

a Mean or $E(X) = 2.0$

b $\Pr(X \leq \mu) = 0.4$

MATCHED EXAMPLE 12

$a = 0.1$ and $b = 0.4$

MATCHED EXAMPLE 13

Mode $= 5$

Median $= 6$

MATCHED EXAMPLE 14

a $E(X) = 2.1$

b $\text{Var}(X) = 1.29$

c $\text{SD}(X) \approx 1.136$

MATCHED EXAMPLE 15

a $E(5X + 1) = 79.5$

b $\text{Var}(5X + 1) = 85$

MATCHED EXAMPLE 16

a $\Pr(X = 2) = 0.261$

b $\Pr(X \geq 2) = 0.841$

MATCHED EXAMPLE 17

a **i** $\Pr(X = 7) = p^7$

    ii $\Pr(X = 6) = 7p^6 - 7p^7$

b $p = \dfrac{7}{8}$

MATCHED EXAMPLE 18

$n = 50$ and $p = \dfrac{2}{5}$

CHAPTER 11

MATCHED EXAMPLE 1

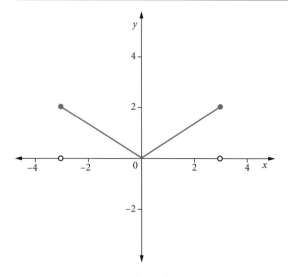

MATCHED EXAMPLE 2

$a = 5$

MATCHED EXAMPLE 3

$$a = \frac{1}{2}$$

MATCHED EXAMPLE 4

$\Pr(X < 4) = \dfrac{45}{56}$

MATCHED EXAMPLE 5

$\Pr\left(X < \dfrac{3}{2}\right) = \dfrac{7}{8}$

MATCHED EXAMPLE 6

$\Pr(T < 80 \mid T \geq 50) = 1 - e^{-20}$

MATCHED EXAMPLE 7

$E(X) = \dfrac{5}{3}$

MATCHED EXAMPLE 8

$E(X) = 1$

MATCHED EXAMPLE 9

$a = 3$

MATCHED EXAMPLE 10

$\mathrm{Var}(X) = \dfrac{5}{63}$

MATCHED EXAMPLE 11

$\Pr(-1.71 \leq Z \leq 1.71) = 0.9128$

MATCHED EXAMPLE 12

$\Pr(X < 36) = 0.0228$

MATCHED EXAMPLE 13

a $\Pr(X < 18) = 0.3446$

b $\Pr(15 < X < 25) = 0.6827$

c $\Pr(X < 18 \mid 15 < X < 25) = 0.2723$

MATCHED EXAMPLE 14

$\sigma = 25.95$

CHAPTER 12

MATCHED EXAMPLE 1

The population is people who visit the Chadstone Shopping Centre.

The minimum amount, maximum amount and mean amount spent at the Chadstone Shopping Centre are parameters. Other parameters are standard deviation, range, mode, median etc.

The statistics are the minimum amount ($25.50), maximum amount ($120.50), range ($95.00) and mean ($55.50).

MATCHED EXAMPLE 2

Proportion related to mathematics ≈ 0.427

MATCHED EXAMPLE 3

The sample proportion of people having wine at home is 0.3.

MATCHED EXAMPLE 4

a The probability is about 0.004 from a sample of 12.

b The probability is about 0.001 08.

MATCHED EXAMPLE 5

$\dfrac{7}{26}$ of the draws are expected to be an ace.

MATCHED EXAMPLE 6

The estimated probability of the people not being artists is about 0.86

MATCHED EXAMPLE 7

0.0810

MATCHED EXAMPLE 8

The probability of 30 to 40 people out of 300 being artists is about 0.52.

MATCHED EXAMPLE 9

The standard deviation of the sampling distribution is about 0.07303.

MATCHED EXAMPLE 10

A sample of about 5200 would be needed.

MATCHED EXAMPLE 11

The 95% confidence interval is $(0.28, 0.62)$

MATCHED EXAMPLE 12

The 95% confidence interval for the germination rate is about $(0.690, 0.810)$.

MATCHED EXAMPLE 13

We would need to ask a minimum of 271 Year 12 students to get a 90% confidence level.

MATCHED EXAMPLE 14

About 34 buses were surveyed.

MATCHED EXAMPLE 15

At least 138 Year 12 students should be surveyed.